13908711792

历史保护建筑图例

■ 文明街小银柜巷 7 号建筑立面测绘图

■ 翠湖南路 4 号建筑实景摄影图

昆明历史建筑

周峰越 主编

YNK 云南科技出版社
·昆明·

图书在版编目（CIP）数据

昆明历史建筑 / 周峰越主编 . -- 昆明 : 云南科技出版社 , 2023.11
ISBN 978-7-5587-5364-0

Ⅰ . ①昆… Ⅱ . ①周… Ⅲ . ①古建筑－建筑艺术－昆明 Ⅳ . ① TU-092.2

中国国家版本馆 CIP 数据核字 (2023) 第 205640 号

昆明历史建筑
KUNMING LISHI JIANZHU
周峰越 主编

出 版 人：温 翔
责任编辑：邓玉婷 王 韬
整体设计：长策文化
责任校对：孙玮贤
责任印制：蒋丽芬

书 号：ISBN 978-7-5587-5364-0
印 刷：昆明亮彩印务有限公司
开 本：787mm × 1092mm 1/16
印 张：9.25
字 数：240 千字
版 次：2023 年 11 月第 1 版
印 次：2023 年 11 月第 1 次印刷
定 价：68.00 元

出版发行：云南科技出版社
地 址：昆明市环城西路 609 号
电 话：0871-64192372

主编

周峰越

副主编

高雪梅　金浩萍

摄影

段　文　刘伶俐

英文翻译

卢雨奇

前 言

关于昆明，虽然在隋朝开皇五年（585年）置昆州，唐代宗永泰元年（765年）南诏在昆州置拓东城，但是，我们最要致敬的是元世祖至元十一年（1274年）在云南的赛典赤·赡思丁先生，他主滇时立云南行中书省于鄯阐，1276年改鄯州为中庆路，昆明县隶属中庆路，成为云南行省的省会所在地。从此以后昆明就一直是云南的省会直至今天。在宝善街与正义路交会处的“忠爱坊”，就是昆明市民对他的敬仰与纪念。

本书所叙述的是自昆明建城以来而又能保存至今的历史建筑，它让我们看到了昆明传统的建筑、街道和村落。一千多年来，昆明城内宫殿学院殿阁云集、名门府邸鳞次栉比；这些优美的亭阁坊表、书院学宫无一不记录着昆明这座城市的悠久历史、灿烂文化的轨迹。在建于唐代的拓东城址之上，我们仍然可以看到清末民初的昆明街道格局和建筑遗存；在滇池东岸的村落，我们仍然保留有贝丘遗址和村舍。这些充满活力的建筑与空间，不仅承载了历史的记忆，还向我们展示了昆明城千年以来的人居环境的变迁。

1982年，昆明被国务院公布为第一批国家历史文化名城。从2002年9月昆明市政府批准第一批33处历史建筑开始，至2023年，昆明已经批准公布了6批历史建筑，共112处建筑。这些能够幸运保留下来的历史建筑是对中国近现代艰难奋斗历史的承载，也是对昆明近现代城市发展的延续。在这本书里，我们看到了这些历史建筑，让我们铭记历史、畅想未来；在这本书里，我们还看到了众多值得保护的建筑与现代建筑和谐共存，让我们深切地感受到这座城市的美丽与魅力。

昆明是一座含蓄、谦逊，然而伟大有力量的城市，在最近一百多年里，辛亥重九起义、护国运动与北伐讨袁、一二·一爱国学生运动、西南联大等都在这座历史文化名城中写下了光彩夺目的历史篇章。相信细心的读者一定会在本书中昆明历史建筑的文脉里，领略到历史篇章意味深长的壮美旋律。

周峰越

2023年10月于昆明理工大学

Preface

About Kunming, was established first as Kunzhou County in Sui Dynasty (585 years) and set up as Tuodong City in Tang Dynasty (765 years). However , the most significant historical event occurred in the Yuan Dynasty eleventh year (1274 years) when Saidianchi-Shansiding played a pivotal role as the first governor in Yunnan Province. In 1276, he made Kunming the capital of Yunnan Province, a status it retains to this day. The "Zhong'ai Fang" Archway at the intersection of Baoshan Street and Zhengyi Road is where the people of Kunming honor and commemorate him.

This book delves into the historical buildings that have been preserved since Kunming's founding, offering a glimpse of the traditional buildings, streets, and villages. For over a millennium, Kunming has been adorned with palaces, academies, and residences. These structures bear witness to the city's long history and splendid cultural trajectory.

Built on the site of Tuodong City of the Tang Dynasty, the street pattern, and architectural remains of Kunming from the late Qing to the early Republic of China era are still visible. Along the east bank of Dianchi Lake, villages retain the ruins of shell mounds and village houses. These vibrant buildings and spaces not only hold the memory of history but also reveal the changes in Kunming City's habitat over the past thousand years.

In 1982, Kunming was declared one of the first batch of National Famous Historical and Cultural Cities by the State Council. Starting in September 2002, when the Kunming City government approved the first batch of 33 historical buildings, and up to 2023, Kunming has endorsed and announced 6 batches of historical buildings, totaling 112. Retaining these historical buildings is a testament to China's modern history and the challenging struggles faced. It also signifies the continuation of modern urban development in Kunming. This book showcases these historical buildings, urging us to remember the past and contemplate the future. It highlights the coexistence of many worthy-of-protection historical buildings and modern architecture, allowing us to deeply feel the beauty and charm of this city.

Kunming is a subtle, modest, yet great and powerful city. In the past hundred years, it has been the backdrop for significant events like the Xinhai Rebellion, the Protectorate Movement and the Northern Expedition against Yuanshikai, the 12·1 Patriotic Student Movement, and the Southwest Association of Universities, all contributing to its glorious historical chapters. Careful readers will appreciate the meaningful and beautiful melody of Kunming's historical buildings portrayed in this book.

Zhou Fengyue

Kunming, Oct. 2023

昆明历史建筑

目录

1 文明街

< 片区 >

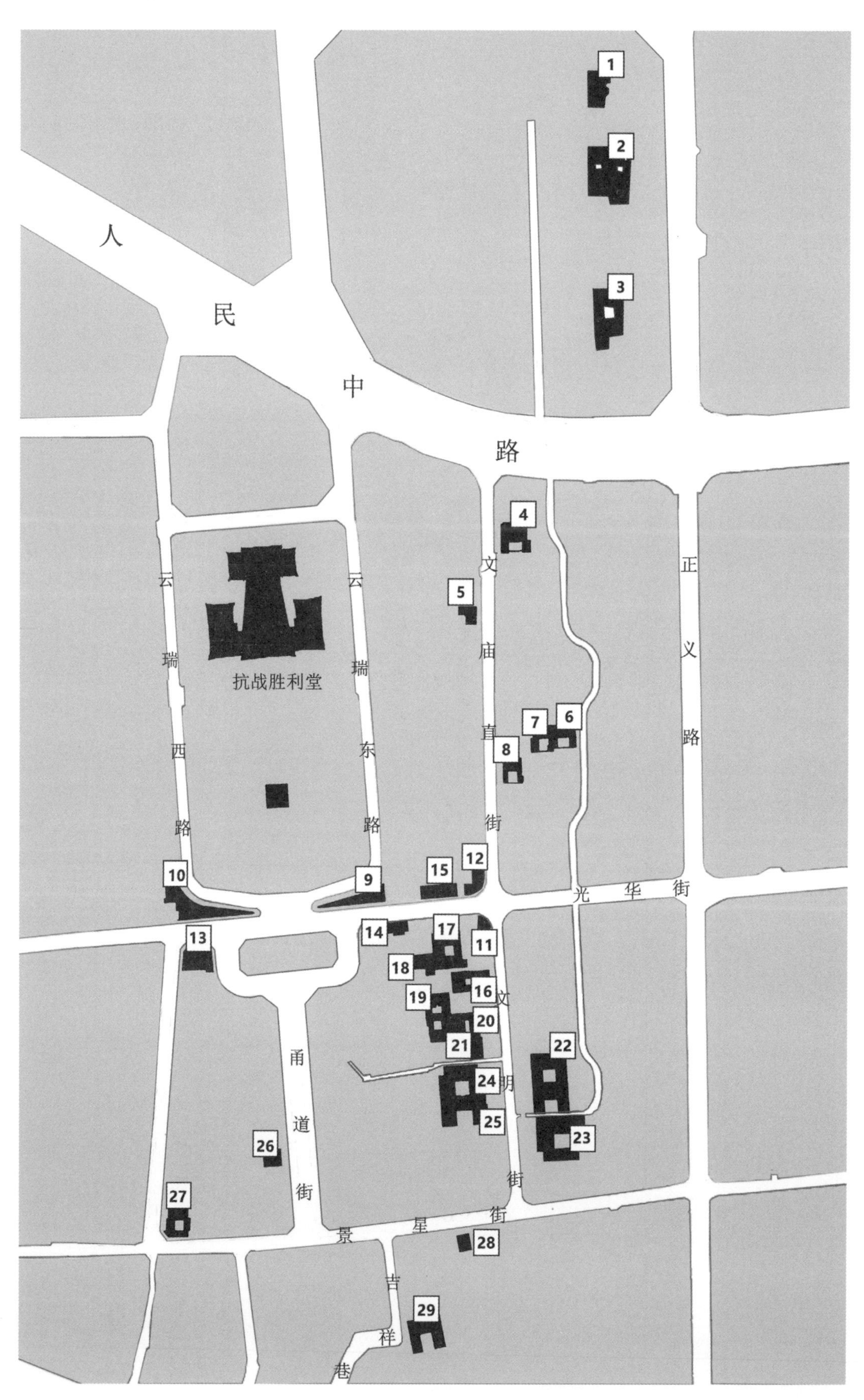
人民中路
云瑞西路
抗战胜利堂
云瑞东路
文庙直街
正义路
光华街
甬道街
文明街
景星街
吉祥巷
1
2
3
4
5
6
7
8
9
10
11
12
13
14
15
16
17
18
19
20
21
22
23
24
25
26
27
28
29

1 文明街

<片区>

区位

1 正义路四通巷 3 号别墅
2 正义路四通巷 2 号宅院
3 藜光庐
4 同庆丰商号旧址
5 文庙直街 78 号宅院
6 居仁巷 11 号宅院
7 傅式宅院
8 居仁巷 8 号宅院
9 酒杯楼（东楼）
10 酒杯楼（西楼）
11 光华街 33 号宅院
12 光华街 38—44 号宅院
13 光华街 63—67 号宅院
14 光华街 45—51 号宅院
15 光华街 46—50 号宅院
16 文明街幸福巷 6 号宅院
17 文明街幸福巷 5 号宅院
18 文明街幸福巷 4 号宅院
19 文明街幸福巷 1 号宅院
20 文明街 28 号宅院
21 文明街 22 号宅院
22 富春恒商号旧址
23 马家大院
24 文明街 16 号宅院
25 欧式宅院
26 甬道街西卷洞巷 1 号宅院
27 景星街 92—94 号宅院
28 景星街 30 号宅院
29 樾庐

1 文明街片区

◈ 历史沿革、保护简况

位于昆明历史城区的文明街历史文化街区是该地区的两个独特的历史文化街区之一。区域北至华山南路，南至东风西路，西至云瑞西路和五一路，东至正义路，总面积为21.92公顷。该街区包含22条历史街巷，如文明街、甬道街、光华街、景星街等。在这个街区中，有一条历史悠久、风貌传统的街道，它不仅建筑质量最高，而且让人流连忘返。1998年，为了保护这片瑰宝，我们特地制定了街区保护规划，并将这个区域正式命名为文明街片区。自此以后，每当提起文明街，人们便会想到这个具有代表性和象征意义的街区。

南诏时期，修筑拓东城就已经将文明街片区纳入城市范围。元代在此修建了昆明文庙和南城清真寺，从而推动了文明街区的繁荣。到了明清时期，随着布政使司署和督抚衙门的设立，行政职能不断加强，形成了今日街坊格局。清朝康熙年间，经过重建的昆明文庙开始接收汉族和少数民族的学生，使得多民族文化得以融合。咸丰年间，文明街成为了集居住、商铺、戏院、茶肆等场所的繁华之地。民国初年，辟粮道署为文明街，而督抚衙门也改建为云瑞中学。虽然作为行政中心的职能逐渐衰落，但商业和文化依旧在这里繁荣发展。1946年，云贵总督衙门旧址上兴建的抗战胜利纪念堂落成。

1983年，《昆明城市总体规划》将文明街历史街区纳入保护范围，1993年开始进行保护规划研究，1996年昆明与苏黎世合作，确定“完整保存历史文化遗产”的目标。2001年成立了“文明街历史街区保护修建指挥部”，2002年至2023年昆明市公布了6批历史建筑，文明街历史文化街区共有29处建筑入选。历次昆明旧城改造中，文明街一直被视为重要的历史文化街区并得以保留。

地名由来

文明街：原为粮道衙门所在地，后于1917年经过改造，成为一条便捷的城市街道。其北端建有南国文明坊，因此得名文明街。

光华街：源自明清时期的东院街，从正义路一直延伸至云瑞公园。由于云瑞公园直达五一路的辕门口，故此段街道被称为辕门口。1911年辛亥革命胜利后，人们以光复中华之意，将其更名为光华街。

景星街：清代光绪年间，粮道衙门设于此地，取名粮道街。1911年辛亥革命的胜利让景星街的街名诞生。传说，当时东方天空出现彩云，西方天空出现景星，作为吉祥的象征。

市府东街：在清代时建有龙王庙，因而被称为龙王街。1927年，它成为国民时期昆明市市政厅所在地，从此便改名为市府东街。

甬道街：清初时期，这里是通向云贵总督府门前的甬道，这就是它名字的来由。

小银柜巷：东起正义路，西至文明街。民国年间，在这里开设有官办银号，为区别于大银柜巷，取名为小银柜巷。

居仁巷：原来的居仁巷起止于正义路西侧，呈“匚”字布局。清代巷内建有土主庙，故称小土主庙巷；民国年间，巷内有居仁里，故改称居仁里巷；1979年将相连的庆鱼巷并入，改称居仁巷；2008年将居仁巷的4处宅院整体西迁20米，现在该巷道西起文庙直街，东至钱王街。

幸福巷：位于云瑞公园东侧，东起文明街北段，西阻。清代以前，这里是官府官吏的私人宅地，1949年后改称幸福巷。

吉祥巷：连接昆明市主干道东风西路、传统老街景星街的一条南北向、曲折蜿蜒的巷子。清代成巷时以有吉祥庵得名吉祥庵巷，辛亥革命后，省去庵字而名吉祥巷。

1.1 Ma Family Compand 23

Ma Family Compand is a traditional residential building. The traditional style green brick tile roofs of Ma Family Compand, brick inlaid Western-style window sets, and carved porch doors, highlight the outstanding and extraordinary of this building during the Republic of China.

The former owner of this house is related to the Ma family in Dali, Yunnan Province. It is the old residence of Ma Jinji and his three sons, so it is called the “Ma Family Compound”.

1.1 马家大院，文明街小银柜巷 7 号 23

文明街马家大院是一座传统民居建筑，传统风格的青砖瓦顶，砖嵌的西式窗套，雕花的连廊门扇，彰显了这座民国时期建筑的卓尔不凡。这座宅院昔日的主人与云南大理的马家有关，是马金墀及其三子的旧居，故被称为“马家大院”。

马家大院位于文明街小银柜巷7号，占地面积约773平方米，建筑面积达1143平方米，为二层土木结构、坐北朝南的传统建筑。由4个条形传统二层建筑围合为中间是大天井、四周有4个小天井的传统院落，因此称为“四合五天井”。出挑的回廊连通二楼所有房间，故称“走马串楼角”。建筑整体方正简洁，大小天井错落有致，房屋檐板门窗木雕的技艺精湛、寓意吉祥平安，抱头梁下有灯笼垂柱，外墙嵌有细格子窗，是昆明老城现存较为完整的传统民居。

【Historical background】

Ma's family was a famous family in Eryuan County, Dali, Yunnan Province. Ma Jinchi was a scholar in the fifth year of Guangxu in the Qing Dynasty (1879). His three sons Ma Zhen, Ma Ying and Ma Yin, alll graduated from Yunnan Military Academy. The eldest son, Ma Zhen, was the most famous. He participated in the Kunming Double Ninth Uprising and defended the country. In 1928, Long Yun, chairman of the National Yunnan Provincial Government, appointed him as the first mayor of Kunming during the Republic of China. The second son, Ying, participated in the Taierzhuang Campaign, the Wuhan Defense Battle, and the Chongyang Battle, and was deputy director of the Yunnan Appeasement Office. The third son, Yin, was a major general and deputy commander of the 11th Division of the Yunnan Army.

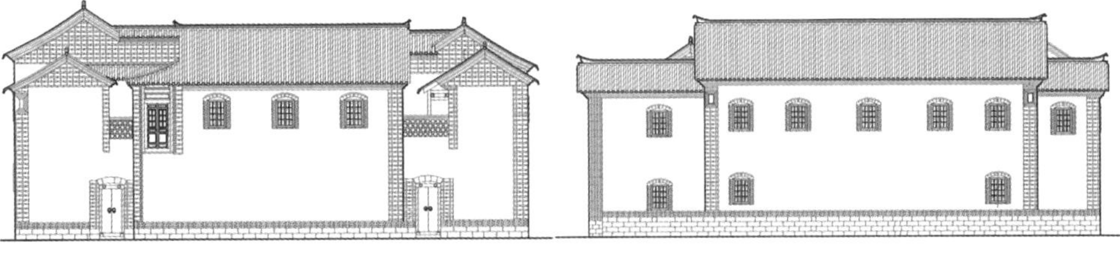

【历史背景】

马家是云南大理洱源县的名门望族，马金墀是清朝清光绪五年（1879年）中的举人，他的三个儿子马钤、马锳、马崟都毕业于云南陆军讲武堂。长子马钤最为出名，他曾参加过昆明重九起义和护国运动，1928年由国民云南省政府主席龙云任命为民国时期昆明市第一任市长，曾任云南省军管区中将副司令。次子马锳曾参加台儿庄战役、武汉保卫战、崇阳会战，曾任云南绥靖公署中将副主任。三子马崟曾任滇军第十一师少将副师长。马家三兄弟故有“一门三将，三迤一家”之称。

1950年该宅院收归国有后作为住房使用，1999年成为金沙地产公司的房产后进行修缮后成为市民休闲的金兰茶苑。2001年该宅院的保护性再利用，获得了2001年联合国教科文组织亚太地区历史遗产保护奖。2007年被昆明之江置业有限公司收购，2010年再次修缮。随着昆明老街的繁荣，现在马家大院成为了昆明小有名气的小剧场，常年在此演出《雷雨》《我的闻先生》等实景话剧，举办形式多样的小型音乐会。

1.2 Ou Shi House 25

In 1927, the house was completed, covering an area of 2,000 square meters and a construction area of 1,400 square meters. There are wells, flower beds and pools in the courtyard, and the homeowners pay attention to the creation of living environment. The house adopts the layout of "three squares and one screen wall". The small wood works are exquisitely carved and the patterns are exquisitely designed. The carving of eaves, birds, hanging columns and doors and windows all reflect the superb construction skills and exquisite materials. At the same time, the house incorporates Western architectural style: French colored tiles adorn the floor of the house, and traditional Chinese lattice doors are inlaid with French embossed glass.

1.2 欧氏宅院，文明街 11 号 25

欧氏宅院是滇军第二军旅长欧阳永昌的私宅。这座宅院从砖墙上的斗拱、青瓦到门前的雕花木质大门，无一不透着精致考究之感。院内布置简洁明快，但每一处都展现着主人的品位和精心经营。同时，庭院中的假山水景、绿植花木更是增添了几分自然之美，让人流连忘返。

欧氏宅院建于1927年，占地面积约2000平方米，建筑面积约1400平方米。建筑为“三坊一照壁”布局，在照壁前方建造有花坛、水池，庭院中设有水井，可见房主十分注重对宅院居住环境的营造。这座宅院的檐坊、雀替、垂柱和门窗等小木作，其图案设计精巧、雕刻精美，建筑用材考究、建造技艺高超。另外，西式建筑风格也融入其中，房屋地面铺装了法式彩色花砖，格扇门镶嵌的玻璃是法国压花玻璃等。

【 **Historical background** 】

In 1927, after Hu Ruoyu, commander of the Second Army of the Yunnan Army, launched the failed coup d 'etat of “Two · Six” in 1927, Ouyang Yongchang built this house in Wenming Street after his resignation. In 1952, Ouyang Yongchang’s younger son, Ouyang Shi Biao, sold the house, and then It was transferred to the Panlong District Cultural Center in Kunming.

Ouyang Yongchang (1887-1942), born in Ouqiying Village, Lin 'an (today’s Jianshui), was a student of the first phase of the Yunnan Military Academy. He participated in the Revolution of 1911 and the National Protection Movement.

【历史背景】

1927年，滇军第二军军长胡若愚发动“二·六”政变失败后，欧阳永昌辞官在文明街建造了这个宅院。1952年，欧阳永昌的幼子欧阳师表出售了宅院，而后转由昆明市盘龙区文化馆使用至今。

欧阳永昌（1887—1942年）字职斋，临安（今建水）欧旗营村人，云南陆军讲武堂一期学生，参加过辛亥革命和护国运动。

1.3 Maolu 29

Maolu (18-19,Jixiang Lane,Jingxing Street) is located on the side of the ancient lane path, covering an area of about 900 square meters, with the construction area of about 700 square meters. It is said that during the Xianfeng Period of the Qing Dynasty, the Kunming squire Zhang Maodi funded the construction of this private house, which was named Maolu.

The main room has five double-pitched roofs, and the two side rooms are three single-eared hard hilltops connected by cloisters, and the gate is located in the southeast corner. On the beams of the building, wood carvings such as "Double Phoenixes Flying towards the Sun" "Double Lions Playing the Hydrangea", It shows a rich cultural atomosphere.

It shows a rich cultural atomosphere.

1.3 懋庐，景星街吉祥巷 18—19 号 29

懋庐坐落于古巷幽径之侧，占地约900平方米，建筑面积约700平方米。据说，在清代咸丰年间，昆明乡绅张懋弟出资兴建了这所私宅，从而得名懋庐。

懋庐是昆明“一颗印”建筑的代表之一，为二层砖木结构的三合院。正房采用五间双坡屋顶，两侧耳房则是三间单檐硬山顶，并通过回廊相连，大门设在东南角。建筑的梁枋上刻有金粉描绘的“双凤朝阳”“双狮戏绣球”“一路（鹭）连（莲）科”等木雕，彰显出浓郁的地方文化氛围。西式砖拱门套的大门，雄健有力的守门石狮，雕梁画栋的金色彩绘，是一处饶有趣味，小而美、小而精的传统建筑。

1.4 藜光庐，文庙东巷 5 号 3

藜光庐位于昆明文庙东巷，与文庙仅一墙之隔。因大门门匾上题有“藜光庐”而得名。藜光庐是二层土木结构的传统合院式建筑，其布局方正紧凑，院落二层通过走廊连通各个房间，门窗梁枋的木雕纹饰精美。

1.5 Tongqingfeng Trade House 4

Tongqingfeng Trade House, located at 103 Wenmiao Straight/Vertical Street, adopts a combination of Chinese-Western courtyard architecture style, with a solid structure using the civil and hard mountain tile roof. The capital under the eaves is finely carved, the railing is hollow-carved and the style is unique, and the doors and windows are ornate and beautiful. It is worth mentioning that the inner courtyard wall of the first floor of the building is different from the wooden structure used in traditional residential buildings, but the use of Western masonry, giving people a sense of simplicity and generosity. At the same time, the stone masonry of the wall skirt also increases the sense of stability of the building, and the white wall plastering is more dignified and elegant.

1.5 同庆丰商号旧址，文庙直街 103 号 4

文庙直街103号曾为王炽创办的同庆丰商号旧址，建筑面积约308平方米，为二进三院二层土木结构硬山瓦顶传统建筑。同庆丰商号旧址是钱王街的标志性建筑。钱王街是2008年文明街历史街区改造提升时所形成的一条新的街道。由于同庆丰是云南钱王王炽所设立的，因此在2008年街区改造建成后，取名为钱王街。

这座位于文庙直街103号的宅院，采用中西合璧的三合院建筑风格，结构坚固，采用土木和硬山瓦顶。其檐下柱头雕刻精细，栏杆镂空雕花样式别致，门窗雕刻华丽，美不胜收。值得一提的是，该建筑一层庭院内墙不同于传统民居采用的木质结构，而是使用西式砖石砌筑，给人古朴大方之感。同时，墙裙的条石砌筑也增加了建筑的稳固感，白墙抹灰更显端庄典雅。正房的大门和窗户更是镶嵌着石膏条的拱形门套和窗套。

歇山屋顶为古代中国建筑屋顶样式之一，它在规格上仅次于庑殿顶，常出现在宫殿、祠庙坛社、寺观衙署等。在同庆丰商号旧址的传统商业建筑中，出现了歇山屋顶的建筑形式，足以证明当时红顶商人王炽具有很高社会地位。

4

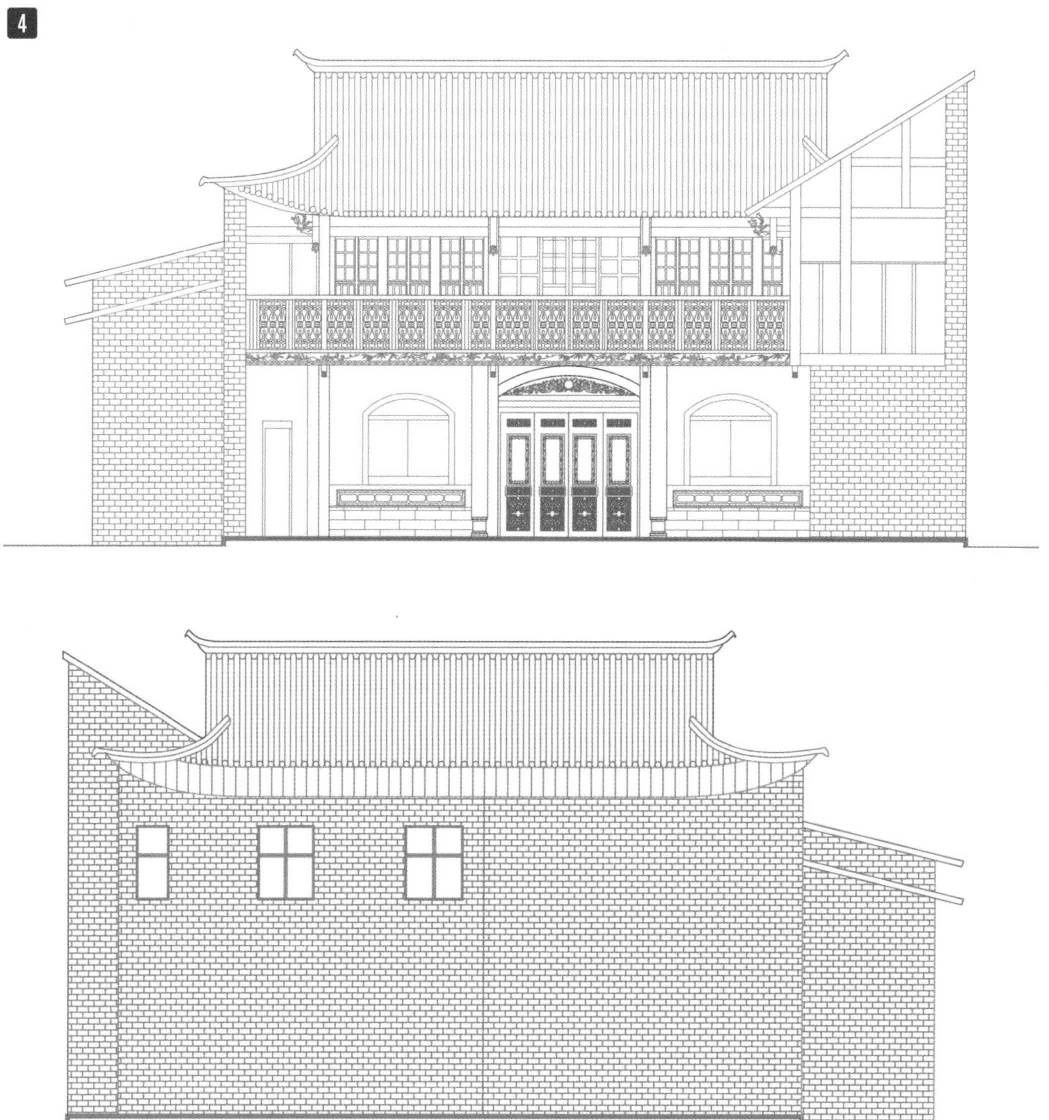

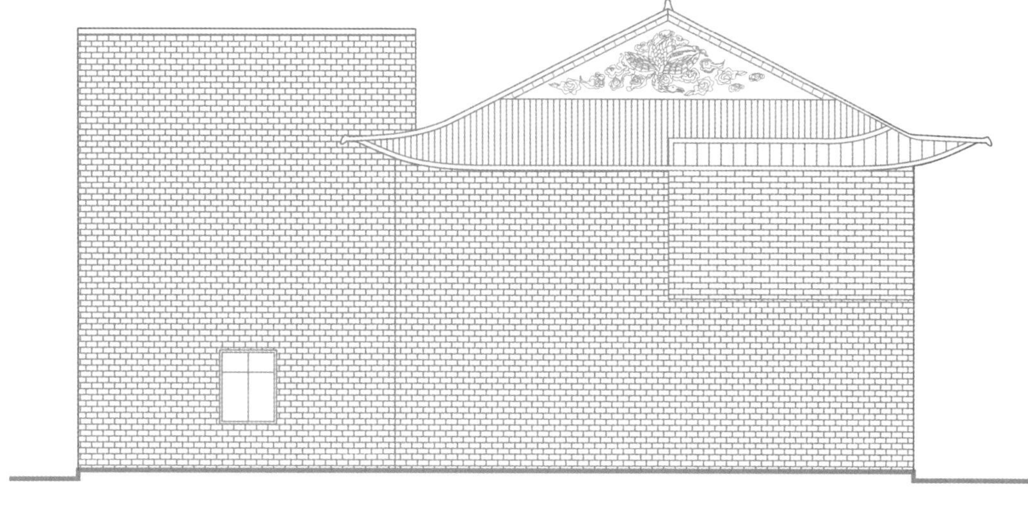

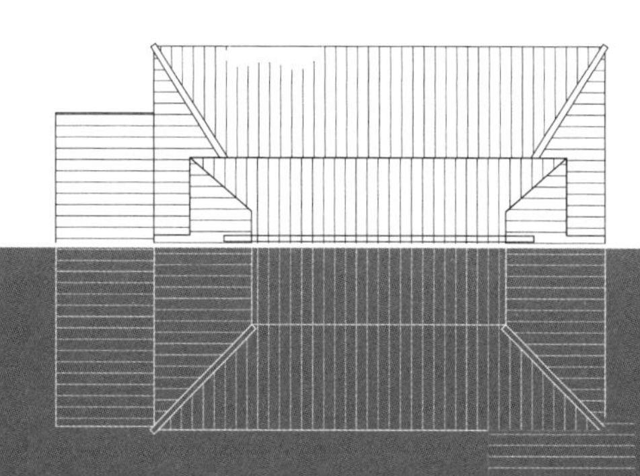

【 **Historical background** 】

Wang Chi (1836-1903), a native of Mile County, Yunnan Province, was a leading merchant of Yunnan in the late Qing Dynasty, and was conferred the title of “Three generations of the First Rank” by the Empress Dowager Cixi.In 1872, Wang Chi set up the “Tongqingfeng” ticket number in Kunming, Kunming “Tongqingfeng” (private bank), the land is in the Qiujia Lane between Nanzheng Street and Wenmiao Straight Street (that is, today’s Qian Wang Street). It began with the “Tianshun Xiang” in Chongqing as the general number, and later promoted the “Tongqingfeng” in Kunming to the general number. In the late Qing Dynasty, Yunnan’s financial funds were kept and transferred by the “Tongqingfeng” ticket number.

【历史背景】

王炽（1836—1903年），云南弥勒人，是清末滇商泰斗，被慈禧太后册封“三代一品”。

1872年，王炽在昆明设立“同庆丰”票号，昆明“同庆丰”(钱庄)，地址在南正街与文庙直街间的邱家巷(即今钱王街)。开始以设在重庆的“天顺祥”为总号，后将昆明的“同庆丰”升为总号。晚清时期，云南财政款项都由“同庆丰”票号保管并承汇。

1.6 Old site of Fuchun heng Business House 22

The former site of Fuchun heng Business House is located on the west side of Qian Wang Street. In 1924, it was built by the descendants of Jiang Zonghan, the founder of Fuchunheng. The south courtyard is the residence of Jiang family, and the north courtyard is the business place of Fuchunheng Business house. The east and west wings of the north courtyard and the first floor of the inverted atrium are connected to the shop fronts of the south courtyard, the warehouse and Man Ming Street respectively. This layout and design not only provides convenience for family-style business operations, but also reflects the cultural connotation.

1.6 富春恒商号旧址，文明街小银柜巷 8 号 22

富春恒商号旧址位于钱王街西侧。1924年由富春恒创始人蒋宗汉的后人所建，南院是蒋家居住，北院为富春恒商号经营场所。北院的东西厢房及倒坐明间一层都作为花厅，分别与南院、库房及文明街的商号门面所连。这种布局和设计，不仅为家族式的商业经营提供了便利，也体现了中式建筑的文化内涵和历史沉淀。富春恒商号旧址是昆明老城中留存不多的南北合院的二进院建筑，是文明街历史文化街区内现存面积最大的、规格最高的民居建筑。建筑的“一颗印”和“三房一照壁”布局具有昆明、大理白族两地传统民居的典型特点，西式的卷曲山花门头砖雕、花砖装饰，这些都体现了昆明历史上多民族文化、中西文化的融合和包容，其建筑风格和设计元素融合了不同民族、不同地域的文化艺术特色，呈现出独特的多元文化氛围，建筑的布局和功能设置也充分考虑了居住、商业的功能需求，体现了昆明地区多元文化的精神内涵。

富春恒商号旧址是昆明市一座具有重要历史价值和文化意义的中式建筑，其建筑风格典雅、古朴，体现了中国传统建筑的精髓和美学价值。

【Historical background】

Jiang Zonghan (1838-1903) was born in Heqing, Yunnan Province. He participated in the suppression of Du Wenxiu's uprising at the end of the Qing Dynasty. In 1876, Jiang Zonghan, Ming Shugong and Dong Yisan opened Fuchunheng in Tengyue. From 1920 to 1930, Fuchunheng's business reached its peak, with more than 40 shops and factories, and the "Lion Ball brand" silk and "Peach brand" flour produced were the epitome of the national industrial development at that time. In 1928, the general number of Fuchunheng Business House was moved to Small silver Cabinet Lane, Wenming Street, Kunming. With the outbreak of the Anti-Japanese War, the economic downturn made the operation increasingly difficult, and in 1937 Fuchunheng went bankrupt.

【历史背景】

蒋宗汉（1838—1903年）是云南鹤庆人，清末参与镇压杜文秀起义，官至参将，腾越厅总兵、贵州提督。1876年蒋宗汉、明树功、董益三在腾越开设福春恒。1920年至1930年间，福春恒的业务达到鼎盛，其商号、工厂达40余处，生产的“狮球牌”细丝、“桃牌”面粉是当时民族工业发展的缩影。1928年，福春恒商号总号迁到昆明文明街小银柜巷。随着抗日战争的爆发，经济低迷，经营日渐困难，1937年福春恒倒闭。1946年后卖给金姓矿主，1948年弃房移居美国旧金山。1950年后该房产作为无主房收归国有，而后作为企业职工宿舍使用。2007年被昆明之江置业有限公司收购。

1.7 House Yard, No.1, Xingfu Lane, Wenming Street 19

No. 1 Xingfu Lane, Wenming Street, was built in the Republic of China period. It is a two-storey building with earth-wood structure. This residence is very distinctive, with a total area of about 800 square meters. The whole building is based on a square plan, composed of square rooms and wing rooms. The main room adopts the design of hard hill with double-slope tile roof, and the continuous "waist building" between the side room forms the upper and lower eaves. In Kunming villages, this kind of "one seal" style of residence with heavy eaves is very common, but in Kunming city, there are few such buildings.

19

1.7 文明街幸福巷 1 号宅院 19

文明街幸福巷1号宅院，建于民国时期，是一个二进院土木结构的二层建筑。这个民居十分有特色，总面积大约800平方米。整个建筑是以方形平面为主，由正房和厢房组成。正房采用双坡瓦顶硬山式设计，与厢房之间的连续“腰厦”构成了上下重檐。在昆明村寨中，这种重檐的“一颗印”形式的民居非常常见，但在昆明城中，很少出现这样的建筑。

然而，这个宅院最引人注目的地方却不是这些传统的设计，二楼过街楼下部分被改造成了宅院的入口。当你从昏暗的过道进入小院的时候，仿佛整个世界都变得明亮起来，让人感到豁然开朗。这样精心的设计，不仅让整个建筑更加具有魅力，也让我们更深刻地了解其历史和文化价值。

1.8 House Yard, No.4, Xingfu Lane, Wenming Street 18

No. 4 Xingfu Lane, a two-storey civil structure, has a traditional architectural style of "three lanes and one view of the wall", with a hard hilltop and a carved and decorated capital eaves, and an ancient well standing quietly in the courtyard. Entering Xingfu Lane, and turning at the old corner, you can see the impressive gate of the house. The wall is made of black brick, showing a calm and simple atmosphere. In particular, the triangular spire above the gate, the brick lines on both sides, and the concave door plaque, all appear solemn and elegant. The brown wooden door complements the arched beams to make the entrance more beautiful. Under the sunlight, the shadow of the door moves with the change of light, creating a very three-dimensional effect.

1.8 文明街幸福巷 4 号宅院 18

幸福巷4号这座二层土木结构的建筑拥有传统的“三坊一照壁”建筑风格，硬山顶和雕刻装饰的柱头檐下，古老的水井在院中静静地伫立。

走进幸福巷，拐过古老的街角，便能看到那所宅院引人注目的大门。墙面由青砖砌成，流露出沉稳、古朴的气息。特别是大门上方三角形的尖顶，两侧的砖砌线条，以及内凹的门匾，都显得庄重而优雅。棕色的木质大门与拱形门梁相得益彰，使整个建筑入口更加美观。在阳光的照射下，大门的影子随着光线的变化而移动，营造出极具立体感的效果。无论是从外部还是内部，这个建筑入口都有着绝佳的视觉效果，让人流连忘返！

1.9 House Yard, No. 5, Xingfu Lane, Wenming Street 17

Located at No. 5 Xingfu Lane, Wenming Street, the mansion is a traditional building built during the Republic of China, adopting the layout of "three squares and one view wall". The courtyard is a two-storey civil structure, with three wide faces and two deep ones, and a roof on the top of the hill. The eaves, capitals, balustrades, doors, windows and doors are decorated with exquisite carvings, presenting a strong historical atmosphere.This house is closely connected with No. 4 Xingfu Lane, continuing the green brick exterior wall, which together constitutes a landscape in Xingfu Lane. From the hustle and bustle of wenming street into the Xingfu Lane, people seem to be in a quiet paradise, feeling the rare quiet and leisure in the busy city.

1.9 文明街幸福巷 5 号宅院 17

位于文明街幸福巷5号的宅院，是一座建于民国时期的传统建筑，采用了“三坊一照壁”的布局。该院落为二层土木结构，面阔三间，进深两间，歇山顶屋面，并采用了走马转角楼的建筑样式。檐下、柱头、栏杆、门窗和门头均有精美的雕刻装饰，呈现出浓郁的历史氛围。

这座宅院与幸福巷4号相连，延续了青砖的外墙面，共同构成了幸福巷中的一道风景线。从喧嚣的文明街步入到幽静的幸福巷中，使人仿佛置身于一个宁静的世外桃源，感受到闹市中难得的宁静与闲适。

1.10 House Yard, No. 6, Xingfu Lane, Wenming Street 16

Fulin Hall is one of the oldest extant pharmacies in Yunnan, founded in 1857 by Mr. Li Yuqing during the Xianfeng period of the Qing Dynasty. Following the example of The Three Kingdoms, he treated the poor people for free, and only asked those who recovered to plant apricot trees in the back hall, where gradually formed a dense apricot forest. Therefore, the drugstore was named "Fulin Hall", meaning "Fuze Xinglin". Located at the corner of Wen Ming Street and Xingfu Lane, the House of No. 6 Xingfu Lane is the pharmacy of Fulin Tang Pharmacy. The building adopts the traditional architectural style of Kunming area, which is an indispensable part of Fulin Hall old shop and has a long history value.

1.10 文明街幸福巷 6 号宅院 16

位于文明街与幸福巷拐角处的幸福巷6号宅院，是福林堂药店的药房。该建筑采用了昆明地区传统建筑风格，是福林堂老店不可或缺的一部分，具有悠久的历史价值。福林堂是云南现存最古老的药店，直至今天还在营业，清朝咸丰年间（1857年）由李玉卿先生创建。他效法三国董奉，为穷苦百姓免费治病，只要求病愈者在后堂植杏树，逐渐形成了茂密的杏林。因此，药店取名“福林堂”，寓意着“福泽杏林”。

幸福巷6号宅院由临街商铺、传统四合院共同组成，呈“前店后院”布局模式。东侧条式建筑为铺面，正房、厢房为居住，正房与厢房保存完好，檐枋、格扇门、花格窗雕刻精美。从北厢房开门可以进入到文明街幸福巷，四合院坐西朝东，砖木结构悬山式屋顶，历经多次修缮保护。

坐落在文明街与光华街街口的福林堂药店，其店与药房仅相距10余米，便于福林堂的经营。

1.11 House Yard, No. 16, Wen Ming Street 24

No. 16 Wenming Street House was built in the period of the Republic of China, near East Juandong Lane. It is connected with the southern side of the European-style house, and the two buildings are three squares and one view wall, which is a rare traditional building community in the historic district of Wenming Street. On the second floor of the house is a corridor connecting the wing room and the main room, and the architrave and lattice doors in the courtyard are carved with auspicious patterns, which are decorated by traditional lacquer gold and tracing technology, which is both antique and show the style. The house is a combination of Chinese and Western elements in some parts, skillfully replacing the traditional lattice windows with the popular Western mosaic glass.

1.11 文明街 16 号宅院 24

文明街16号宅院建于民国时期，靠近东卷洞巷。它与欧式宅院南侧相连，两处建筑都是“三坊一照壁”式宅院，这在文明街历史街区中是罕见的传统建筑群落。

宅院的二楼由走廊贯穿连接厢房和正房，天井内的挑檐枋和格扇门都雕刻有吉祥图案，通过传统漆金和描金工艺进行粉饰，既古香古色，又彰显气派。该宅院在局部中西合璧，巧妙地将当时流行的西式镶嵌玻璃替代了传统花格窗，这些细节体现了在民国年间昆明民居建筑多元文化融合的特点。

1.12 文明街 22 号宅院 21

这座宅院位于文明街22号，建于民国时期。它是一座二层土木结构硬山顶屋面的建筑，院落狭长，内部装饰朴素。底层面向街道的部分用作商铺，而二层则作为居住区域。这种布局方式符合传统的“下店上住”式商住建筑的设计风格。

1.13 No. 28, Wenming Street 20

Located at No. 28 Wenming Street, the house was built at the end of the Qing Dynasty, and after many repairs, it still retains the characteristics of traditional Qing architecture. The two-storey civil structure of the building, using the hanging mountain tile roof, is small and exquisite, and its column head, door and window carving and balustrade decoration are also unique.In addition, it is one of the historic buildings of Wen ming Street, fronting the street and having a narrow passageway connecting with the courtyard behind the storefront. The pool in front of the wall adds a bit of elegance to the house, and it is a beautiful small yard worthy of fine taste.

1.13 文明街 28 号宅院 20

位于文明街28号的宅院，建于清末，历经多次修缮，仍然保留着清代传统建筑的特色。该建筑二层土木结构，采用悬山式瓦屋顶和走马转角廊，小巧而精致，其柱头、门窗雕花和栏杆装饰也别具一格。

因用地有限，建筑布局狭长，包括一个“三坊一照壁”和一个四合院，两个院落之间通过下空的过街楼相连。此外，它是文明街历史建筑之一，临街铺面并设有一狭窄通道与店面后的合院相通。照壁前的水池为宅院增添了几分雅致，是一处值得细品的精美小院。

1.14 No. 1 West Juandong Lane, Yongdao Street 26

No. 1 West Juandong Lane, Yongdao Street is a traditional residence built in the period of the Republic of China, located in Yongdao Street. It used to be a pizza shop, but now is used a jewelry shop. The house adopts the typical "three square and one illuminated wall" style, which is a two-storey building with civil structure. There are large and small patios in front and back, paved with stones, making people feel as if they were on a mountain path.

1.14 甬道街西卷洞巷 1 号宅院 26

西卷洞巷1号宅院，是一座建于民国时期的传统民居，坐落于甬道街。曾用作比萨饼店，现为珠宝店。该宅院采用典型的“三坊一照壁”样式，为土木结构两层建筑。前后各有一大、一小天井，院中铺有条石。硬山瓦顶、走马转角楼，檐下柱头、门窗、垂花等小木作雕刻精致、装饰华丽。

1.15 正义路四通巷 2 号宅院 2

这座位于正义路四通巷2号的宅院，建于民国时期，采用纵向布局，结构主要以土木为主。它由北院的二层一颗印式建筑和南院的一层东西耳房组成，北院建筑的二层通过木质走廊相互连接。小木作雕饰细腻典雅，增添了别致的传统民居气息，令人心旷神怡。

1.16 正义路四通巷 3 号别墅 1

位于昆明文庙东侧的正义路四通巷3号别墅，建于民国时期。这栋别墅呈半圆形和“L”形不对称组合，坐西向东，是一座二层砖木结构建筑，灰瓦屋面和青砖墙面使其散发出近现代建筑风格的简洁与考究。

1.17 Fu Family House 7

In the Fu Family House, the Chinese-style Kunming "One seal" style is based, and Western architectural elements are cleverly integrated into the design of the building railings and terraces. The courtyard is made of civil structure, which is surrounded by the main room, the east wing room, the west wing room and the inverted seat. There is a corridor running through the whole courtyard on the second floor. The main house has three floors and three eaves on top of the hill. East and west wing of the second floor flat roof hanging eaves, the second floor roof for the terrace, the south of the terrace built east and west a Feiling corner pavilion, corner pavilion four wings of the spire, the top of the treasure reversely-set room.

1.17 傅式宅院，居仁巷 10 号 7

居仁巷10号宅院是1931年由云南火腿商人傅润之、傅泰之兄弟二人及父亲所建，故称为傅氏宅院。宅院中小巧玲珑的东、西两个角亭尤为引人注目，使之成为文明街街区中最为漂亮、最为别致的一处宅院。这处建筑不仅记录了傅氏家族的发展与兴衰，也是当时昆明民居建筑技艺水平的历史见证。

在傅氏宅院中，中式昆明“一颗印”样式为基础，西式建筑元素巧妙地融入栏杆和露台的设计中。宅院采用土木结构，由正房、东厢房、西厢房和倒座合围而成，二层有回廊贯穿整座院落。正房为三层重檐歇山顶；东、西厢房的二层平顶挂檐，二层屋面辟为露台，露台南端东、西各建一处飞棱角亭，角亭四面翼角攒尖顶，上置宝顶。宅院的围栏由整齐排列的陶制宝瓶柱和条石组成，倒座的走廊则装饰有弧形拱券和罗马柱式。建筑以中式风格为主，但在门柱、雕刻、纹饰等装点了欧式元素，彰显着其奢华和非凡之处。西角亭于20世纪70年代被拆毁，2008年得以复建。

7

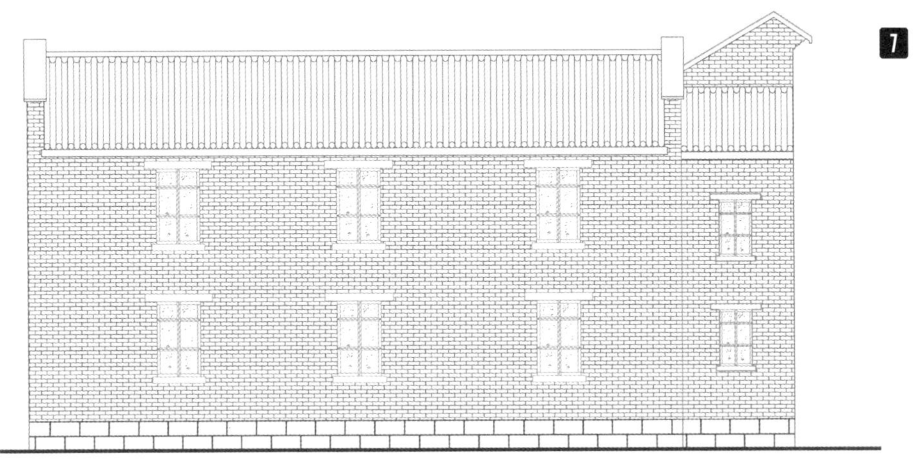
7

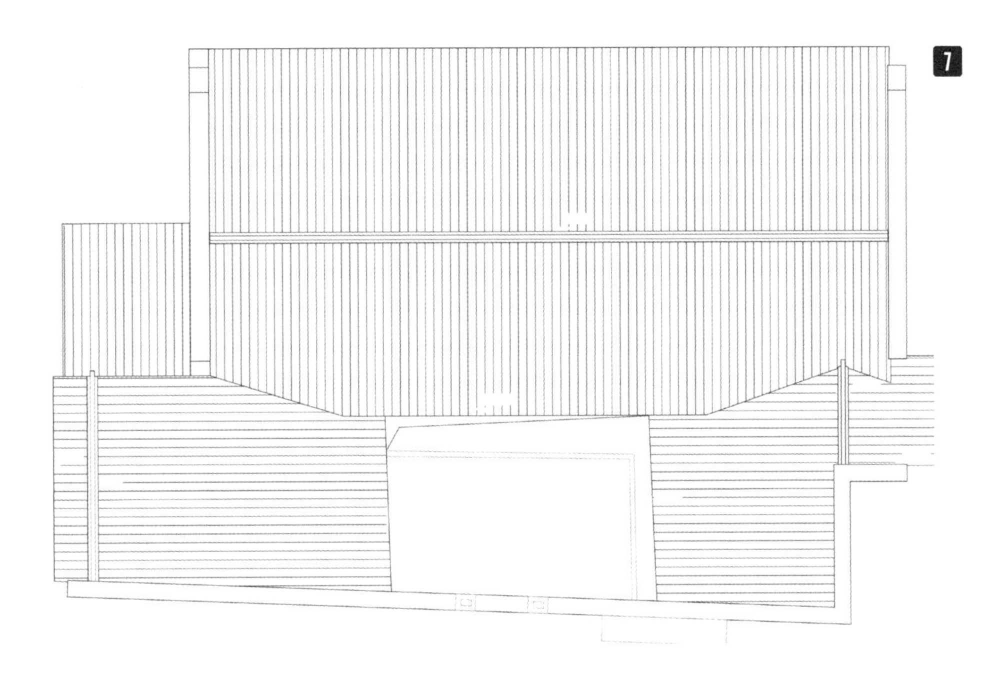
7

【Historical background】

The Fu family originated from Zhanyi Couty of Qujing Region. When they were young, the parents came to Weiyuan Street Kunming to help relatives with business. After business experience, they opened their own independent shop. Thanks to Zhanyi's proximity to Xuanwei, Xuanwei is famous for producing high-quality ham. So Fu famliy decided to open a shop specializing in selling ham, and it was an instant success. Subsequently, he invested in the construction of a luxury house in Juren Lane, where rich people gathered. When the house was completed, the ham business began to decline sharply, and Fu was on the verge of bankruptcy. Fu's house also experienced the war years in that era.

7

【历史背景】

傅氏家族起源于曲靖沾益，傅润之、傅泰之的父母年轻时到昆明威远街帮衬亲戚生意，有了经商经验后，便自己独立开店。靠着沾益与宣威较近，而宣威以出产优质火腿闻名。于是，他决定开一家专门卖火腿的店铺，并迅速取得了成功。随后，他在富人聚集的居仁巷投资建造了豪华的宅院。当宅院建好之时，火腿生意开始急剧下滑，傅氏濒临破产。傅氏宅院也与那个时代共同经历了烽火岁月。

在红军长征入滇后，傅氏资产连同宅院一并被收缴充公；1950年公私合营后，该宅院成为原昆明市糖业烟酒公司职工宿舍；2006年房屋产权转至昆明之江置业有限公司。

1.18 No. 8, Juren Lane 8

The second floor of No. 8 Juren Lane house adopts civil structure. The top is a hard hill, and the overall building is in the style of "Three squares and one view wall". The building's brick and wood carvings take a variety of forms, including phoenixes, bat-patterned balustrades, and lantern-like pendents. In addition, the door also uses the brick carving of Western architectural elements, such as spires and mountain flowers. The whole courtyard layout is compact and unique, which fully shows the characteristics of Kunming residential buildings in the early Republic of China.

1.18 居仁巷 8 号宅院 8

居仁巷8号宅院的二层采用土木结构，顶部为硬山顶，整体建筑呈“三坊一照壁”式样。建筑的砖木雕饰形式多样，包括凤头造型的雀替、蝙蝠图案的栏杆以及灯笼式垂柱。此外，门头还采用了西式建筑元素的砖雕，如尖顶饰和山花等。整个院落布局紧凑、别具一格，充分展现了民国初年昆明民居建筑的特色。

1.20 No. 92-94 Jingxing Street 27

No. 92-94 Jingxing Street is located at the intersection of Jingxing Street and East Shifu Street. As a representative of the traditional "cight-sided wind" front-shop and back house style, its complete layout and the two-storey brick and wood building on the hard hill constitute the quadrangle coutyard. The building serves as a node building at the corner of the street entrance, and the curved street facade and other traditional buildings outline the continuation of the historical district.

1.19 居仁巷 11 号宅院 6

建于民国时期的居仁巷11号宅院，建筑形式为“三坊一照壁”，二层土木结构，东厢房硬山单坡屋顶，西厢房硬山双坡屋顶。外墙有着砖砌的窗套和门套，弧形砖拱券，尖顶门头和双柱拱券。门格扇、外廊檐枋下多作花牙子。正房二楼的阁楼采用歇山戗角屋顶设计，显得庄重大气，使得这座院落与相邻的傅氏宅院相比也毫不逊色。

1.20 景星街 92—94 号宅院 27

景星街92—94号宅院位于景星街与市府东街的交叉路口，作为传统“八面风”前店后宅式传统建筑代表，其完整的布局和硬山顶的砖木二层建筑构成了四合院。该建筑作为街道口转角上的节点建筑，弧形沿街立面与其他传统建筑勾勒风貌延续的历史街区。

1.21 No. 30 Jingxing Street 28

The courtyard is one of the well-preserved traditional buildings, consisting of three two-storey brick and wood buildings in the south, north and west, with a traditional front shop and back house design. As a street building, this house and other traditional buildings formed a continuous style of traditional commercial district. It is a representative of the type of traditional commercial buildings along the street in Kunming.

1.21 景星街 30 号宅院 28

位于景星街30号宅院是保存完好的传统建筑之一，由南、北、西三栋两层砖木结构的建筑围合而成，采用了传统的前店后宅设计和硬山顶屋顶。作为沿街建筑，这座宅院与其他传统建筑形成了一个具有延续性的传统商业街区风貌。虽然这处院落没有精美的雕饰和严整的布局，但依然是昆明传统沿街商业建筑这一类型的代表。

1.22 文庙直街 78 号宅院 5

文庙直街78号宅院由“一字型”建筑和坐西朝东的“一颗印”式建筑构成，二层砖木结构，前店后宅。每个建筑构件都别具匠心，雕刻精湛，为昆明传统街巷历史留下了重要的记忆。

1.23, 1.24　Wine Glass Building 9

In front of the Kunming Anti-Japanese War Victory Memorial Hall, there are two curved buildings, which are built symmetrically along East Yunrui Road and West Road, known as the "Wine Glass Building". The designer of the two buildings was Li Hua of Tsinghua University. It was organized by the Kunming Municipal Public Works Bureau, and Shanghai Lugen Ji construction factory is responsible for the construction. The East Building of "goblet Building" (No. 74-84 Guanghua Street, No. 1 East Yunrui Road) and the West building of "goblet Building" (No. 86-108 Guanghua Street, No. 1-29 West Yunrui Road) are modern and modern public buildings with three-storey brick and wood structure.

1.23　酒杯楼（东楼）9

1.24　酒杯楼（西楼）10

在昆明抗战胜利纪念堂前，伫立着两座弧形建筑，这对姊妹楼分别沿着云瑞东路和西路对称而建，被誉为“酒杯楼”。这两座建筑的设计师是清华大学李华，承办方是昆明市工务局，并由上海陆根记营造厂负责建设。“酒杯楼”东楼（光华街74—84号、云瑞东路1号）与“酒杯楼”西楼（光华街86—108号、云瑞西路1—29号）为三层砖木结构的近现代建筑。两处建筑巧妙地顺应地形，设计为弧形建筑平面、底层内部北高南低。女儿墙的高恰到好处遮挡住了坡顶，为这座看似平顶建筑增添了几分神秘。淡黄色的抹灰外墙与整齐竖立的窗户和横线条装饰相得益彰，使得酒杯楼更显简约大气。而这种设计也更加凸显出胜利堂的庄严气息。酒杯楼的一层商铺，二层和三层分别作为办公区和起居室。然而，经过60余年的沧桑岁月，这两座楼已被时代所遗忘，成为危楼。在2008年，通过保留外墙并加固结构的方式进行全面维修后，如今的酒杯楼已成为老昆明的标志性建筑，它代表着20世纪40年代建筑设计的最高水平，也是那个时代的代表性建筑。

【 **Historical background** 】

On July 7, 1945, at the twenty-first session of the second Yunnan Provincial Provisional Council, it was suggested that the Kunming Government should build a “Zhi Court” in Yunrui Park (the former site of the Yungui Governor’s Office in the Qing Dynasty), but Long Yun thought it was inappropriate and called it "Zhongshan Hall". After the completion of Zhongshan Hall in January 1946, in order to commemorate the victory of the Anti-Japanese War, it should be called the "Victory Hall of the Anti-Japanese War".The two sister buildings are curved like the sides of wine glasses, and the Yunrui Park is its perfect support. The road extending from Guanghua Street to Jingxing Street forms a Western-style goblet building, and the curved section on both sides of Yunrui Park becomes the wall of the cup,

【**历史背景**】

1945年7月7日，云南省临时参议会在第二届第二十一次会议上提议，昆明市政府在云瑞公园（原清代云贵总督府旧址）建设“志公堂”，时任云南省务委员会主席的龙云改名为“中山堂”。1946年1月中山堂建成后，为了纪念抗战胜利，又将其改称为“抗战胜利堂”。云瑞东路、云瑞西路和光华街，构成了一个如同中式酒杯的造型，曲线美妙。两座姊妹楼呈弧形，好似酒杯的杯壁，而云瑞公园则是其完美支撑。而从光华街延伸至景星街的道路，则形成了一个西式高脚酒杯，云瑞公园两旁的曲线路段成为了杯壁，甬道街是杯柄，景星街则成为了它的底座。以中式酒杯和西式高脚杯为灵感，该建筑布局设计将胜利堂屹立于两杯之间，象征着“中轴线上叠双杯，举酒双杯庆胜利”的精神内涵。

1.25 No. 63-67, Guanghua Street 13

No. 63-67 Guanghua Street and Wine Glass building across the street, is a three-storey brick and wood structure of modern public buildings. Together with Yunrui Park and Yongdao Street, it forms the wall of the Western-style tall cup. This semi-arc "eight-sided wind" building, yellow plastered exterior wall, gray and red washed-stone horizontal lines, vertical decorative columns, brown wooden lattice windows and other elements complement each other, which constitutes a modern architectural style, echoing the architectural style of the goblet building.

1.25 光华街 63—67 号宅院 13

光华街63—67号宅院与酒杯楼隔街相望，是一座三层砖木结构的近现代公共建筑。它与云瑞公园、甬道街构成了西式高脚杯的杯壁。这座半弧形“八面风”建筑，黄色抹灰的外墙面，灰红相间水洗石的横线条、竖向装饰柱，棕色木格窗等元素相得益彰，所构成的近现代建筑风格，与酒杯姊妹楼的建筑风格相呼应。

这座建筑是构成“双杯庆胜利”街巷格局的重要组成部分，也是城市发展的记忆留存。

1.26 No. 33, Guanghua Street 11

1.27 No. 38-44, Guanghua Street 12

The east-west Guanghua Street intersects with the north-south Wenming Street and Wenmiao Street. The south side of Guanghua Street is Wenming Street, and the north side is Straight Wenmiao Street. Here you can see that No. 33 Guanghua Street, No. 38-44 Guanghua Street, Fulin Hall (No. 33 Guanghua Street) and other four buildings stand at the intersection of the street, which together constitute an important node of the traditional street. These buildings have a consistent style, curved plan, sloping tile roof, and adopt the traditional eight-sided architectural style. This group of buildings is not only an important part of the memory and impression of Kunming's traditional streets, but also a unique feature in the historical and cultural streets of Wenming Street.

1.26 光华街 33 号宅院 11

1.27 光华街 38—44 号宅院 12

东西向的光华街和南北向的文明街以及文庙直街相交。光华街南侧是文明街，北侧是文庙直街。这里可以看到光华街33号、光华街38—44号、福林堂（光华街33号）等4处建筑在街道交叉口矗立，共同构成了传统街区的重要节点。这些建筑风格一致，弧形平面，坡面瓦顶，采用传统的“八面风”建筑风格。这组建筑不仅是昆明传统街区记忆与印象的重要组成部分，而且在文明街历史文化街区中独具特色。

1.28 No. 45-51, Guanghua Street 14

1.29 No. 46-50, Guanghua Street 15

No. 45-51 and No. 46-50 of Guanghua Street are well-preserved and representative traditional commercial buildings. Brick and wood structure, hard hilltop, shop on the house, the ground floor is used as a shop, the second and third floors are for living. The wooden carvings of the eaves are exquisite, and the narrow lattice windows along the street are simple and orderly. The three-storey loft is located in the center, giving it an imposing presence in a building dominated by two-storey shops on Guang hua.

1.28 光华街 45—51 号宅院 14

1.29 光华街 46—50 号宅院 15

光华街45—51号、46—50号是保存较为完好、具有代表性的传统商业建筑。砖木结构、硬山顶，下店上宅，底层作为商铺使用，二层和三层则供居住之用。檐口的木雕精致，沿街细格窗简约有序。三层的阁楼位于中央，使其在光华街以二层商铺为主的建筑中显得饶有气势。

光华街的建筑为清末民初时期建造，其建筑规整、风格统一，是文明街传统建筑群落中不可或缺的组成部分，是昆明商埠开放、城市发展的重要历史见证。

2 南强街

<片区>

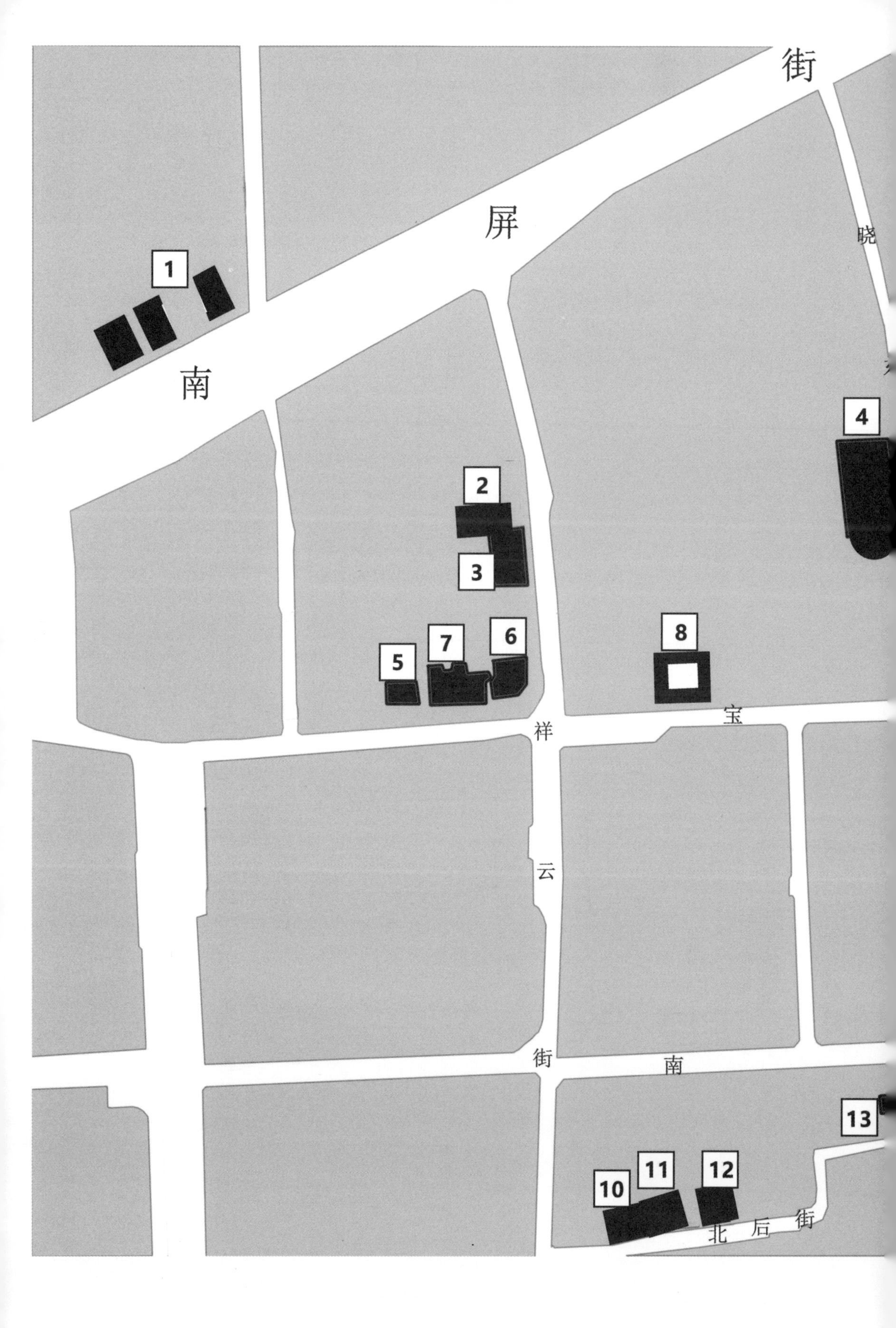

南屏街
晓
祥云街
宝
南
北后街
1
2
3
4
5
6
7
8
10
11
12
13

2 南强街

<片区>

区位

1 南屏街 68—75 号

2 祥云街 43—44 号宅院

3 祥云街 45—48 号宅院

4 南屏电影院

5 宝善街 171—173 号宅院

6 飞虎队俱乐部

7 宝善街 175—177 号宅院

8 宝善街 193 号宅院

9 中国科学院昆明分院办公楼

10 北后街 27—30 号宅院

11 北后街 31—33 号宅院

12 北后街 34 号宅院

13 北后街端仁巷 17 号宅院

14 北后街端仁巷 18 号宅院

2 南强街片区

◈ 地名由来

南强街

南强街历史文化街区，是昆明历史城区中的两个历史文化街区之一，街区面积5.5公顷，北起宝善街，南至端仁巷、金碧路，西至同仁街，东至护国路。

在清初，南强街一带原本是驻军练武、阅兵和行刑示众的南校场。在清代中期，这里逐渐成为珠宝、玉器、毡子、木材和竹器等行业的聚集地。1905年昆明自辟商埠开放后，南强街更成为浙、广商会的汇聚点，1926年南强街逐步形成。民国时期，按照《1922年昆明市政计划大纲》，昆明市开始有计划地整修街道，开发城南区域。在此背景下，南校街进行拓宽改造和沿街建筑整修。

自2013年起，对南强街、北后街和端仁巷一带的传统建筑进行了大规模修缮和修复，使之焕发出新的生机与活力。如今，南强街已成为昆明最受欢迎的传统商业步行街区，吸引着众多游客和市民前来感受这座城市的历史和文化底蕴。因为靠近南校场，故被称为南校街。后为唤醒民众的自强不息精神，南校街被改名为南强街。

南屏街

东起护国路，西至近日楼，南屏街东与东风东路相通，西与东风西路相连。1928年昆明市政府成立后，拆除了正义路以东、护国路以西的城墙，以城墙土填护城河，修筑了宽12米的块石路面，三合土的人行道，原称为新市场。修建南屏街的位置是昆明老城城墙外的护城河，因为街道在城南，在当时是横贯昆明的宽敞平直的东西向要道，故后定名为南屏街，即城南屏障之意。1965年曾改称东风路，1982年又复名。

北后街

北后街的历史可以追溯到民国初年，以房屋建筑大多面向北而得名。

祥云街

南起金碧路，北至南屏街，中跨宝善街。1926年街道形成，取名为祥云街。1947年前后北段形成，称为新祥云街，1979年后南北合称为现在的名称。自20世纪80年代以来，祥云街一带便成为了昆明城中繁华的市井文化娱乐之地，南屏、新昆明、红旗等多家电影院和北京饭店、祥云旅社、东来顺饭店、金碧啤酒店等知名场所坐落其中。

宝善街

宝善街西起正义路，东至盘龙江，与尚义街相连。清代街道的东头有一座石桥，桥上为珠宝交易市场，称珠市桥，又称珠市街。

2.1 No.27-30 North Back Street 10

No. 27-30 North Back Street House, located at the middle section of North Back Street, is a traditional building along the street. Its pattern is complete, "four-horse cart" style, three-room quadrangle courtyard, two-storey civil structure, and an attic is located in the middle of the second floor. This attic has a roof at the hip angle of the mountain, making it the tallest and most distinctive residential building in the Nanqiang Street historic district.

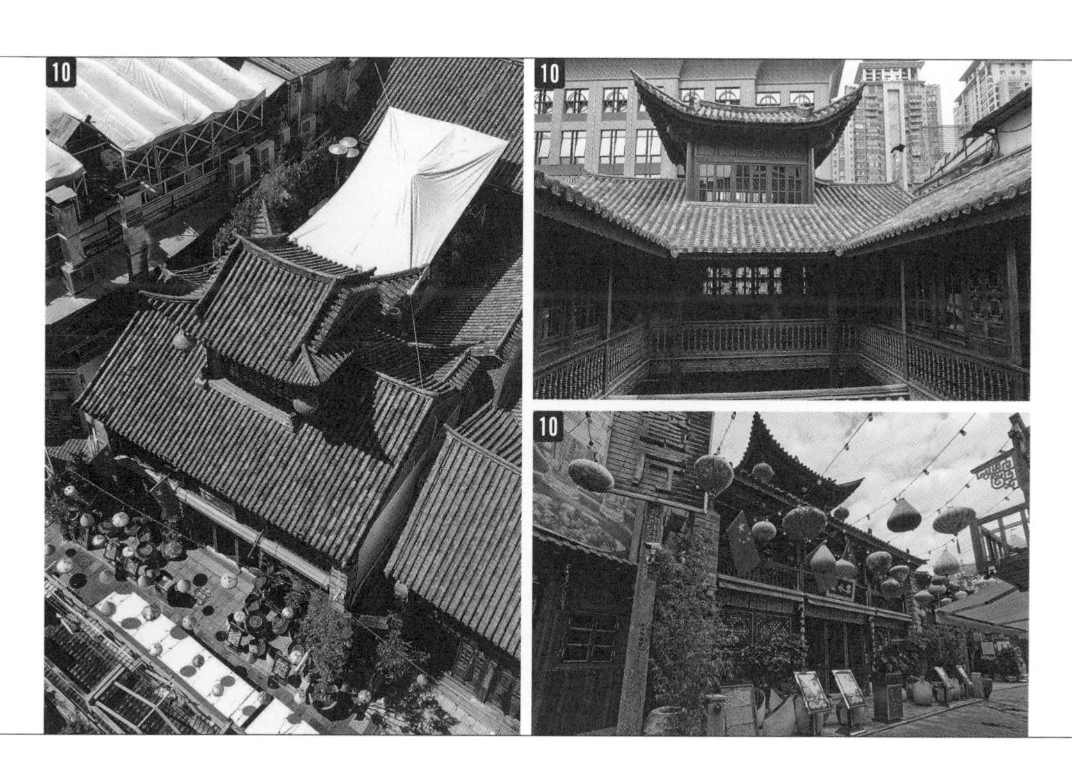

2.1 北后街 27—30 号宅院 10

北后街27—30号宅院坐落于北后街中段，是一座沿街的传统建筑。它的格局完整，"四马推车"样式，三开间四合院，二层土木结构，倒座三层正中设有阁楼。这座阁楼采用歇山戗角屋顶，使其成为南强街历史街区中最高、最具特色的民居建筑。

走进宅院，狭长的庭院映入眼帘。照壁横立于天井中央，将庭院一分为二，游人只能从照壁两侧拱门穿过。照壁前龙头石雕的水景和角落的古朴石井，为闹市中的庭院营造出了一份难得的清雅。

2.2 No. 31-33 North Back Street 11

No. 31-33 North Back Street built in the Republic of China period, it adopts the style of Kunming folk houses. It is a two-storey civil structure building with exquisite carving on the capital, doors and windows under the eaves on the top of hard mountain tiles, full of artistic atmosphere.

The house consists of three courtyards of different sizes, each consisting of three squares. Along the street, two smaller south courtyards and a larger north courtyard are arranged in a " 品 " shape, with three courtyards of different sizes. The south courtyard has a "waist building", which has obvious architectural elements of Kunming's "One seal" residential buildings, showing the unique architectural style and cultural connotation of Kunming residential buildings.

2.2 北后街 31—33 号宅院 11

北后街31—33号宅院建于民国时期，采用昆明民居风格，是一座二层土木结构的建筑，硬山瓦顶上檐下的柱头、门窗、镂空栏杆等雕刻精致，充满了艺术气息。

宅院由三个不同规模的院落构成，每个院落由三坊组成。沿街两个较小的南院与一个较大的北院呈“品”字形布局，三个院落分别设有大小不同的三个天井。南院设有“腰厦”，具有明显的昆明“一颗印”民居的建筑元素，展现出昆明民居独特的建筑风格和文化内涵。

2.3 No. 34 North Back Street 12

North Back Street No. 34 House courtyard, located at the east end of North Back Street and connected with the camp gate, is a traditional courtyard building. The two-storey civil structure and the hard-tiled roof show its ancient traces. Despite years of disrepair (the main room is seriously damaged, and the side rooms on both sides have collapsed), from the building's pattern, capitals, doors and windows, and the carving of railings, you can still glimpse the historical value of this building. After the renovation and renewal of the block in 2003, the house has been completely restored, radiating the extraordinary atmosphere when it was first built.

12

13

14

2.3 北后街 34 号宅院 12

北后街34号宅院位于北后街东端与营门口相连，是一座传统的四合院建筑。二层土木结构和硬山瓦顶展现出其古老之踪迹。尽管经年失修，正房残损严重，两侧厢房坍塌，但从建筑的格局、柱头、门窗、栏杆的雕刻中，仍能够窥见这座建筑所具备的历史价值。在2003年街区整治更新后，宅院得到了完整的修复，焕发出初建时的不凡气息。

2.4 北后街端仁巷 17 号宅院 13

北后街17号宅院是一座硬山瓦顶的二层砖木结构建筑，整体呈横向布局，尺度较小，天井狭长，其民国初年的昆明民居建筑特征在小木作雕刻精细中展现。虽然建筑外观已经破损严重，原沿街隔扇木门也被青砖砌堵，但风貌仍然犹存。2022年，该宅院将进行全面修缮，以完整恢复其传统的沿街商业建筑风貌。

2.5 Office Building, Kunming Branch of Chinese Academy of Sciences 9

Office Building of Kunming Branch of Chinese Academy of Sciences is located at No. 22 Huguo Road, next to the former YMCA site on the west side. It was used as the office building of Yunnan Daily in 1953. The building is located west to east, four-storey masonry structure, blue brick and white paint powder to separate, simple and elegant shape, showing the modern office building characteristics of the times. It can be said that this building is the historical witness of the Chinese Academy of Sciences in Kunming.

2.5 中国科学院昆明分院办公楼 9

中国科学院昆明分院办公楼位于护国路22号，西侧紧邻基督教青年会旧址，曾在1953年作为云南日报的办公楼使用。建筑坐西向东，四层砖石结构，建筑立面以水洗石、青砖与白漆粉面进行分隔，造型朴素典雅，展现出近现代办公建筑的时代特征。可以说，这座建筑是中国科学院在昆明办公的历史见证。1957年，中国科学院成立中国科学院昆明办事处，即中国科学院昆明分院前身。1958年扩建为中国科学院云南分院，1962年与四川分院合并，设中国科学院西南分院。1979年经国务院批准成立中国科学院昆明分院。下辖研究所，分设物理、天文、化学、植物、动物、金属、地质、冶金等学科。

2.6 北后街端仁巷 18 号宅院 14

北后街端仁巷18号是一座传统的“三坊一照壁”院落，它的二层建筑采用砖木结构，木作雕刻精细，厢房则为三坡瓦顶，屋顶层次丰富，错落有致。虽然沿街立面传统风貌已被破坏，东厢房也已经坍塌，但这座建筑仍然展现出传统建筑的独特魅力。在2022年重新恢复了东厢房和大门。

2.7 Nanping Cinema 4

Nanping Cinema, also known as Nanping Grand Theatre, is located at No. 2 Xiaodong Street. The elegant building was designed by the renowned architect Zhao Chen and built by Shanghai Lugenji Construction Factory. The social activist and female entrepreneur Liu Shuqing invited Long Yun's wife Gu Yingqiu and Lu Han's wife Long Zeqing to communicate and finance, in 1939, the construction of the South Screen Cinema in the Xiaodong Street started. So it is also known as the "Lady Group" cinema.

2.7 南屏电影院 4

南屏电影院，又称南屏大戏院，位于晓东街2号。这座优雅的建筑由著名建筑师赵琛设计，上海陆根记营造厂承建。由社会活动家、女企业家刘淑清（中华人民共和国成立后，为云南省民建副主任）邀请龙云夫人顾映秋和卢汉夫人龙泽清共同出资，1939年在晓东街修建南屏电影院，因此也被称为“夫人集团”电影院。

这座电影院占地1820平方米，建筑面积1200平方米。观影厅有1400个座位，楼厅有352个座位，座位排列坡度为倒仰式，观众视线与银幕的角度设计得十分精妙。设计风格别具一格，空间布局灵活多变，建筑造型优美动人。从平面组合、立面处理到地形利用，无不展现出独特的设计思想。前厅采用半圆形平面，与观影大厅的弧形屋顶相呼应，让人感觉舒适自然。而轻盈的玻璃窗和稳重的外墙形成了虚实对比，观影大厅外墙带形窗和竖向标牌墙则形成了横纵对比，让整座建筑尽显匠心独运之妙。

南屏电影院是昆明市近代建筑史上的重要代表之一，历经60余年仍保持着原有的风貌。该建筑在建成后曾辉煌一时，被誉为西南地区最出色的电影院之一，甚至媲美当时全国最高级别的南京“大华电影院”和上海“大光明电影院”。在抗战期间，它以电影艺术推动抗战宣传，在西南及东南亚产生了广泛影响。经过1991年的重点抗震加固维修，整个建筑依旧保持着较高的历史价值。

2.8 No. 68-75 Nanping Street 1

No. 68-75 Nanping Street was a representative financial building in the 1940s. The building is 4 to 5 storeys high, with a symmetrical design. The external walls of bean sand stone or washed stone, brown wooden lattice doors and windows, vertical or horizontal lines, masonry and brick-concrete structure, make the architectural style simple and modern.

2.8 南屏街 68—75 号 1

南屏街68—75号是20世纪40年代的代表性金融建筑，建筑为4—5层高，呈对称式设计，豆沙石水洗石的外墙，棕色的木格门窗，装饰竖向或横向的线条，砖石、砖混结构，建筑风格简洁、现代。

1

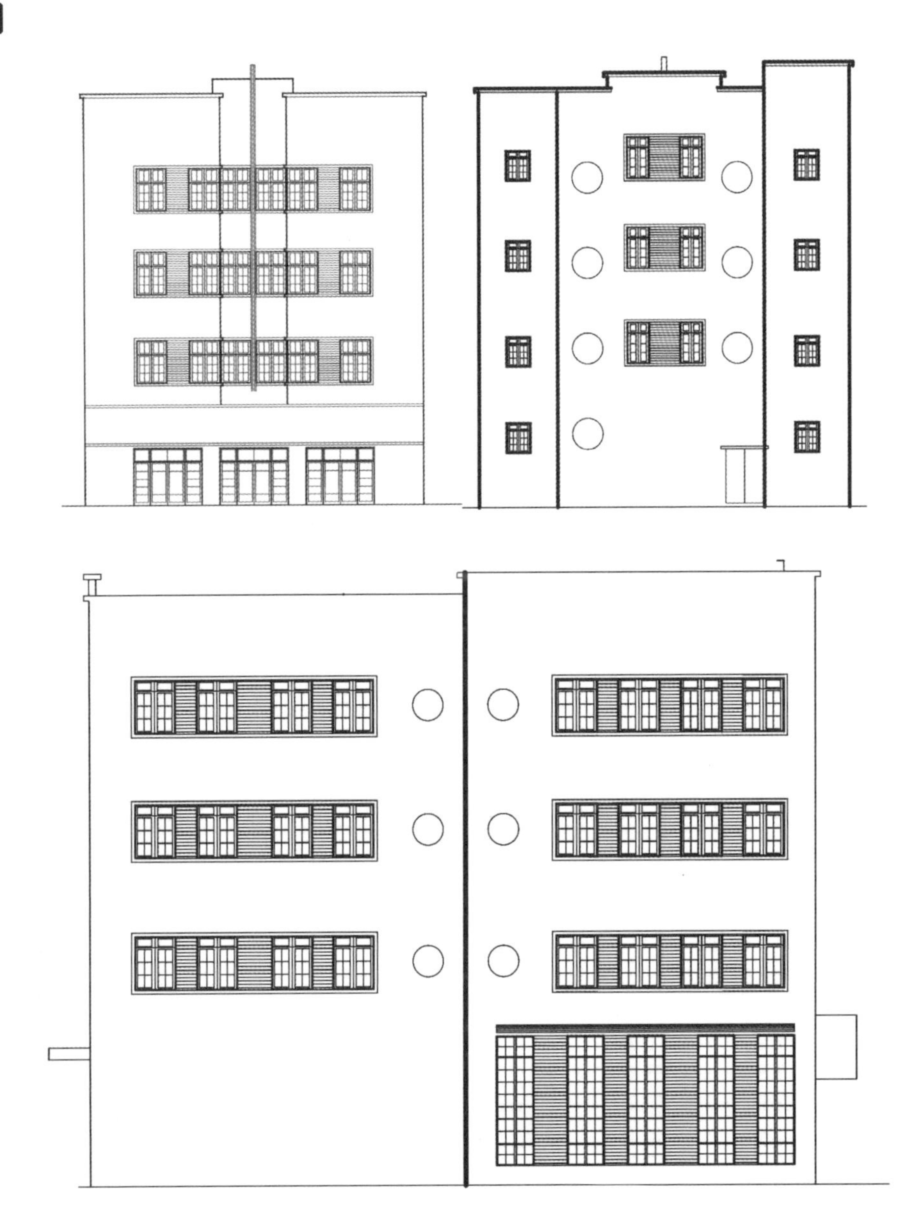

No. 43-44 Xiangyun Street 2 2

No. 45-48 Xiangyun Street 3 3

No. 43-44 Xiangyun Street and No. 45-48 Xiangyun Street are two connected buildings. The fronting buildings of Xiangyun Street and Baoshan Street are influenced by Western architectural styles and most of them are modern in appearance. The interior of the buildings is a " 回 -shaped" layout in the style of traditional Chinese dwellings.

2.9 祥云街 43—44 号宅院 2

2.10 祥云街 45—48 号宅院 3

祥云街43—44号、祥云街45—48号为相连的两处建筑，祥云街、宝善街的临街建筑受到西方建筑风格的影响，外观多为近现代风格，建筑内部仍然是中国传统民居样式“回字形”平面布局，每处建筑中部各有一个庭院，三至五层砖木结构，外立面为浅黄色抹灰，对称式设计，装饰有简约的横向线条。

2.11 No. 171-173 Baoshan Street 5

No. 171-173 Baoshan Street was built between 1940 and 1945. This two-stage courtyard consists of the South Courtyard and the North Courtyard, which are dignified and symmetrical. The main building of the South Courtyard along the street is three storeys high, and the two wing rooms are elegant two-storey structures. The building plane adopts the traditional residential courtyard layout, but the facade and inner courtyard of western modern architectural style are carefully integrated into the architectural design. The south courtyard building along the street displays a sleek modern style with pale yellow exterior walls and brown-red wood windows. The courtyard corridor is not confined to the traditional carved wooden railings, but replaced by brick arches and exquisite hollowed-out brick railings integrating Western architectural elements, which is rare in the residential buildings in Kunming.

2.11 宝善街171—173号宅院 5

宝善街17—173号建于1940年至1945年，这座二进院由南院三合院和北院四合院组成，端庄而对称。沿街的南院主房楼高三层，两厢房则是优雅的二层结构。建筑平面采用了传统的民居三合院布局，但在建筑设计中精心融入了西方现代建筑风格的外立面和内部庭院。沿街的南院建筑展示着浅黄色外墙和棕红色木窗的现代风格。庭院走廊不拘泥于传统的雕花木栏杆，取而代之的是融入西方建筑元素的砖拱券和精致镂空的砖砌栏杆，这在昆明的民居建筑中可谓稀有之物。

北院是一座经典的二层传统民居。从南院的天井穿过花厅，迈进北院，两个庭院之间的走廊巧妙地将二层房间相连。

2.12 No. 175-177, Baoshan Street 7

No. 175-177 Baoshan Street is a commercial building along the street, with a three-storey layout. The main house along the street is covered with a hard tiled roof, while the side rooms have a flat roof design. Although the plane layout shows the characteristics of traditional houses, from the aspects of the facade, inner courtyard and corridor, it uses black brick as the masonry material, and the roof tile is no longer the common blue copper tile of Kunming houses, but changed into red Western-style tile.

The building is a representative example of the transition from traditional Chinese style to Western modern architecture, clearly demonstrating the deep influence of Western culture on Kunming architecture in the 1940s.

2.12 宝善街 175—177 号宅院 7

宝善街175—177号宅院是一座沿街商业建筑，采用了三合院的布局形式，共三层。沿街正房覆盖着硬山瓦顶，而两侧厢房则采用了平顶设计。尽管平面布局上呈现出传统民居的特点，但从外立面、内庭院和走廊等方面来看，却使用了青砖作为砌筑材料，并且屋顶瓦面也不再采用昆明民居常见的青色铜板瓦，而改成了红色的西式瓦。

这座建筑展示了中国传统风格向西式现代建筑转型的代表实例，清晰地表明了20世纪40年代昆明建筑受到西方文化的深刻影响。

2.13 宝善街 193 号宅院 8

位于宝善街193号的宅院是一座建筑是“四合三天井”式的合院。二层土木结构、坚实硬山顶屋面、“走马转角楼”内廊，上层为住宅，下层则设有商铺。整体来看，室内装饰虽然朴素，却彰显出一种简约之美。

2.14 Flying Tigers Club 6

No. 179 Baoshan Street is located at the intersection of Baoshan Street and Xiangyun Street. It is a semi-curved building, five stories high (partially six stories), brick and wood structure flat roof, and one of the important public buildings in Kunming in the 1940s. The symmetrical design of the building is high in the middle and low on both sides. The walls between the Windows with shaped Windows in the southeast and the middle are dark brown bean sand stone, which forms a contrast with the gray washed stone wall. In addition, the horizontal and longitudinal lines of the building facade are decorated, showing the unique charm of "the architectural style of decorative art".

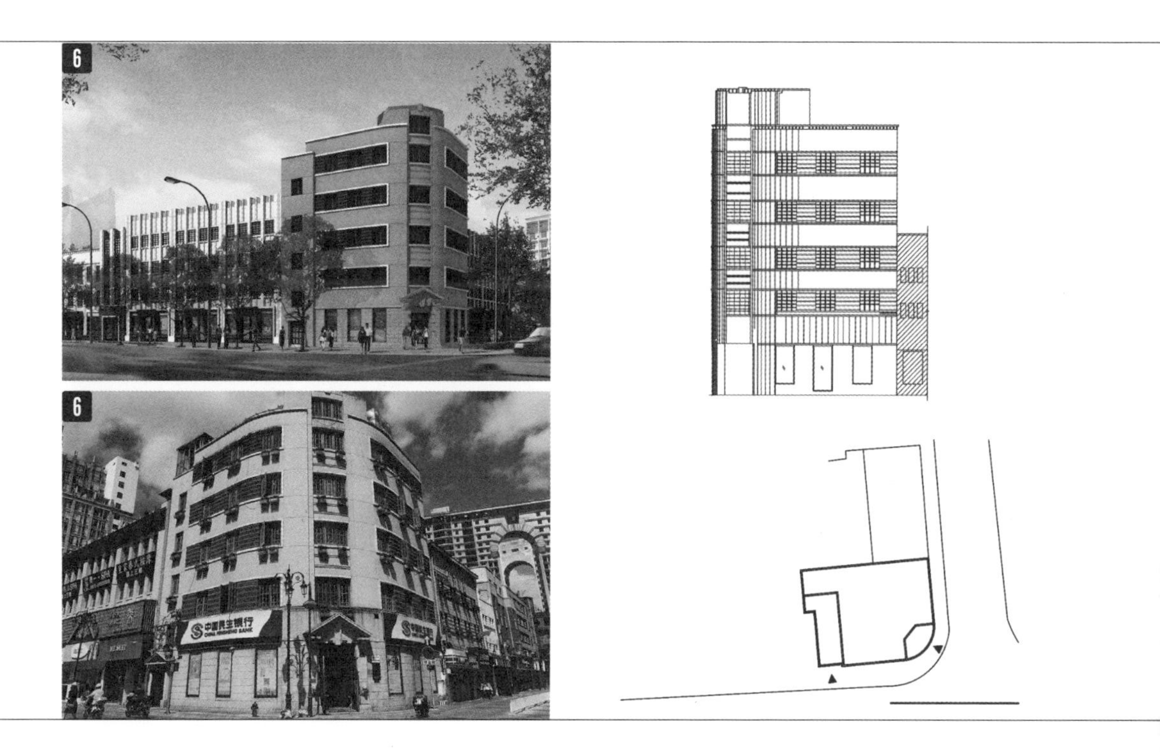

2.14 飞虎队俱乐部 6

宝善街179号位于宝善街与祥云街的交汇处。它是一座半弧形建筑，五层高（局部六层），砖木结构平屋顶，是昆明20世纪40年代的重要公共建筑之一。建筑对称设计，中间高两侧低，东南两面带型窗的窗间墙为深棕色豆砂石，与灰色水洗石墙面形成了明暗对比，加之建筑立面横纵线条的装点，展现出"装饰艺术建筑风格"的独特魅力。

2011年，对宝善街179号进行了保护性修缮。黄色涂料被清除，水洗石材质得以恢复，原有结构部分也得到了加固处理。

宝善街179号的历史可追溯到1945年8月，由昆明颇负盛名的云南保山籍旅缅侨商舒子烈、舒子杰兄弟投资所建。正值抗日战争的日本投降日期（1945年8月15日）所在的8月份，因此该建筑被命名为"国庆大楼"。原本计划用于扩大富滇银行业务，但后来为了感谢美国飞虎队为抗日战争所做出的贡献，将国庆大楼出租给陈纳德作为民用航空售票处及办公场所。它内部布局合理，装修精美，楼下为商业用房，楼上为办公用房。

3
翠湖
< 片区 >

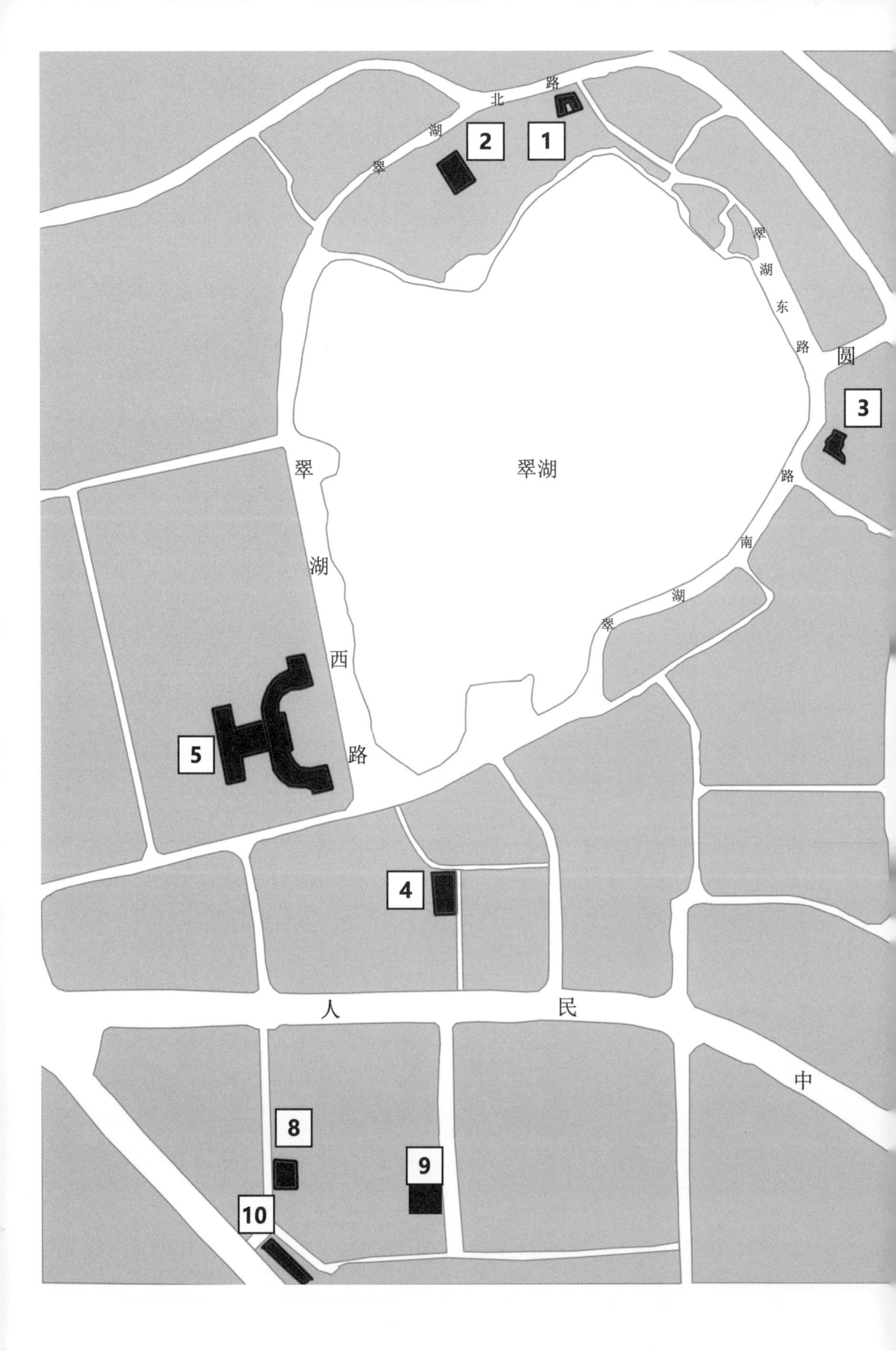
翠湖北路
2
1
翠湖东路
圆
3
翠湖
翠湖西路
路
翠湖南
5
4
人民中
8
9
10

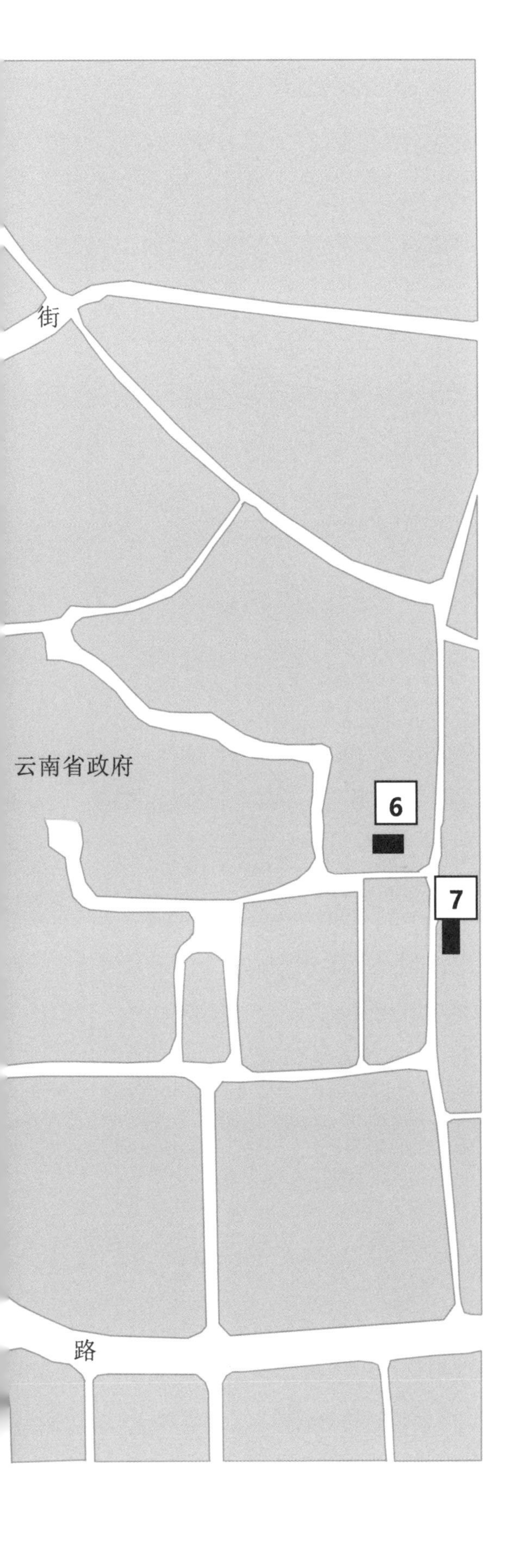

<片区>

区位

1 王九龄旧居

2 袁嘉谷旧居

3 卢汉公馆

4 石屏会馆

5 云南省农业展览馆

6 黄河巷杨氏公馆

7 东苑别墅

8 大富春街何氏宅院

9 富春街 3 号宅院

10 核工业商务楼

3 翠湖片区

◈ 翠湖的历史

因“九泉所出，汇而成湖”古称“九龙池”，又因满池莲荷，周边是菜园，故又称莲湖、菜海子。到民国八年（1919年），改辟为公园，园内遍植柳树，湖内多种荷花，改称为翠湖。

唐代、元代时期，滇池北岸线退至今篆塘一带，螺峰山及五华山的泉水不断，形成湖湾。那时的翠湖仍然是为拓东城、中庆城（今昆明老城）城外的湖泊。明代洪武十五年（1382年），城址北移西扩，将翠湖围入城内，构成“三山一水”的山水格局。沐英在此，设柳营屯兵。清康熙三十一年（1692年），巡抚王继文在湖中小洲建碧漪亭、来爽楼，自此翠湖开放供人游览。清道光十五年（1835年），总督阮元在湖中修筑南北长堤，称“阮堤”，1919年唐继尧筑东西长堤，称“唐堤”，两堤在湖心交汇，将湖一分为四。

◈ 地名由来

华山东路

五华山位于昆明历史城区内，是云南省政府所在地。围绕五华山有3条路，即华山南路、华山西路、华山东路。

在清代，华山东路的南段称为四吉堆，北段因有永宁宫，称永宁坡，在1937年后改成了现在的名称。

富春街

明末清初，江南的一些商贾士绅纷纷迁往云南府城，在这里购买土地，建造房屋。随着时间的推移，这里逐渐形成了一条街道，人们便以故乡的富春江为名，将这条街道称之为富春街。

3.1 Yuan Jiagu's Old Residence 2

Yuan Jiagu's former residence is located at No. 51 North Cuihu Road, built in 1920. When Yuan Jiagu was a professor at Yunnan University, he and his family lived here for a long time. In the early 1950s, the former residence was bought back by Yunnan University.

The building is located at the bank of Cuihu Lake. The courtyard house of the two courtyards, in addition to the main room for the three-storey resting peak, the rest of the buildings for the two-storey hard peak, the overall style is very beautiful. The main room of the house, also known as "Sleeping Snow Hall", is five rooms wide, and the attic on the top floor is Yuan Jiagu's study. The two wings of the house are three wide rooms and five inverted enes.

3.1 袁嘉谷旧居 2

袁嘉谷旧居位于翠湖北路51号，建于1920年。在袁嘉谷担任云南大学教授期间，他和家人长期居住在此。20世纪50年代初，该故居被云南大学购回。

该建筑位于翠湖湖畔，二进院走马转角楼的四合院，除正房为三层歇山顶外，其余建筑为两层硬山顶，整体风貌非常优美。宅院的正房又称“卧雪堂”，宽五间，顶层的阁楼是袁嘉谷的书斋。宅院的两个耳房面阔三间，倒座五间。

【Historical background】

Mr. Yuan Jiagu (1872-1937), addressed respectfully as Shuwu and nicknamed by himself as Shupu, was born in Shiping County, Yunnan Province. He was the first in the first class of economic special science in the late Qing Dynasty and the only champion in Yunnan. In 1904, he was sent to Japan to study, and later returned to serve as director of the Department of Compilation and Books. Subsequently, he served as Zhejiang learning minister and Yunnan salt transport minister. After the outbreak of the 1911 Revolution, he left Zhejiang and returned to Yunnan, serving successively as a member of the provincial administration, a professor at Donglu University and deputy director of the Library of Yunnan Province. In 1923, at the beginning of the establishment of Donglu University (now Yunnan University), Yuan Jiagu was hired as a lecturer in Chinese studies, and he was one of the well-known and widely respected professors of Yunnan University.

【历史背景】

袁嘉谷先生（1872—1937年），字树五，号澍圃，是云南石屏县人。他是清朝末年经济特科一等第一名，也是云南唯一的状元。1904年，他公派赴日本考察，后来回国担任学部编译图书局局长。随后，他担任浙江提学使和云南盐运使等职。1911年辛亥革命爆发后，他离开浙江回到云南，历任省务委员、东陆大学教授和云南省图书馆副馆长。1923年，东陆大学（现云南大学）创校之初，袁嘉谷受聘为国学讲师，他是云南大学广受尊敬的知名教授之一。

袁嘉谷先生著有多本著作，包括《卧学堂文集》《卧学堂诗集》《滇绎》《滇诗丛录》和《石屏县志》《东游日记》四卷等。

3.2 Wang Jiuling's Old Residence 1

The light yellow traditional residence on the shore of Cuihu Lake is the old residence of Wang Jiuling (No. 3 North Cuihu Road). This house was built in 1925, and Mr. Wang Jiuling and his family had lived here for a long time.

Located on the north side of the Cuihu Lake, Wang Jiuling's old traditional residence is a very distinctive building. It is a courtyard building with a "waist building". The building is a masterpiece of ingenuity, carefully designed to be able to better appreciate the view of the lake, reflecting the high quality of architectural aesthetics.

3.2 王九龄旧居 1

位于翠湖湖滨的浅黄色传统民居是王九龄旧居（翠湖北路3号）。这所宅院建于1925年，王九龄先生与家人长期居住于此。

位于翠湖北侧的王九龄旧居传统民居中一处极具特色的建筑。为设有“腰厦”的“走马转角楼”合院式建筑。该建筑是匠心独运的杰作，精心设计以能够更好地欣赏翠湖景观为目的，体现了高品质的建筑美学。

考虑到翠湖在宅院的南侧，建筑设计体现了深度思考和良好的完成度。南侧倒座仅有一层高，二层连廊连接正房和东、西两个厢房。该宅院通过令人心旷神怡的转角阳台设计，将连廊延伸至两厢房的端头，形成了别样的观景方式。与之相呼应的是，厢房的屋顶在面向翠湖一侧增加了坡面，形成了独特的三坡屋顶，这充分体现了建筑对翠湖等自然环境的敬意。

【 Historical background 】

Wang Jiuling (1880-1951), a native of Yunlong County, Yunnan Province, served as director general of the Education Department of the Government of the Republic of China in Beijing, financial director of Yunnan Province, and the first honorary president of Donglu University. In 1937, Donglu University rented the house for professors to live in. It was purchased by Yunnan University in 1952. In 1953, Mr. Liu Wendian, a famous master and professor of Chinese studies, established Du Fu Research Office here.

【历史背景】

王九龄（1880—1951年），云南云龙县人，曾任北京民国政府教育司总长、云南财政司长、东陆大学首任名誉校长。1937年东陆大学租下此宅供教授居住。1952年由云南大学购买。1953年著名国学大师、教授刘文典先生在此设立杜甫研究室。

3.3 Luhan Mansion 3

Located at No. 4 South Cuihu Road, the Luhan Residence, consisting of the old residence and the new residence, was built in the 1930s as the private residence of the former chairman of Yunnan Province, Lu Han, and is now used as the Yunnan Uprising Exhibition Hall.

Luhan Mansion, a French-style building, is displayed like a painting. The stone pillars, door covers, window covers and other stone components are carved with exquisite stone designs, delicate and gorgeous. The building is divided into two floors, brick wood and concrete mixed structure. The geometric composition of the carefully designed facade presents a simple and bright aesthetic.

3.3 卢汉公馆 3

位于翠湖南路4号的卢汉公馆，由老公馆和新公馆组成，建于20世纪30年代，为原云南省主席卢汉私宅，现为云南起义展览馆使用。

卢汉公馆，一座法式风格的建筑，犹如一幅绘画般展现在眼前。石柱、门套、窗套等石构件上雕刻着精美的图案，细腻而华丽。建筑共分两层，砖木、混凝土混合结构。精心设计的外立面的几何构图，呈现出简洁明快的美感。

步入起居室，壁炉散发着温暖的火光，石柱、门窗与窗台上的浮雕装饰展现出精湛的建筑工艺。主次卧室和餐室通过落地式门窗与宽敞的阳台相连，自然光线洒入室内，营造出宜人的居住氛围。东西侧的三面体凸窗，将室内外的景色完美地融为一体。红瓦白墙面相得益彰，灰色边框线条勾勒出建筑的轮廓，整体呈现出和谐而美观的景象。

【 Historical background 】

On December 9, 1949, during the Kunming uprising, Lu Han detained Li Mi, commander of the Eighth Army of the KMT, Yu Chengwan, commander of the 26th Army, and Shen Zui, chief of the Yunnan Station of Juntong. Luhan Mansion is a witness to the major historical event of the peaceful uprising in Kunming, which has high historical value.

【历史背景】

1949年12月9日昆明起义时，卢汉将国民党第八军军长李弥、第二十六军军长余程万、军统云南站站长沈醉等人扣押在此。卢汉公馆是昆明和平起义这一重大历史事件的见证地，具有较高的历史价值。

3.4 Shiping Guild Hall 4

Shiping Guild Hall is located in Zhonghe Lane, South Cuihu Road. Built in 1908, it faces south and backs north. It is a typical residential building of southern Yunnan style, showing a civil structure of three courtyards.

To enter Shiping Guild Hall, you must climb the stone ladder on the right side of the elevation at the end of the lane. The three-section arc wall highlights the grandeur of the building. The layout of the whole house is rigorous and orderly, each courtyard is closely connected, and the second floor courtyard is connected with each other through the corridor.

3.4 石屏会馆 4

石屏会馆坐落于翠湖南路中和巷，建于1908年，面朝南方，背靠北方，是一座典型的滇南建筑风格的民居，呈现出三进院、四合五天井的土木结构。

踏入石屏会馆，须从巷子尽头高地右侧的石梯拾级而上。三段式的弧形围墙，彰显着建筑的庄严、威仪。整个宅院的布局严谨有序，各个院落之间紧密相连，而二层的院落则通过走廊相互串联。

会馆的装饰精妙绝伦，充满着巧思和精湛的技艺。格子门上雕刻着精美的花纹，玲珑剔透的窗户透射出柔和的光线。建筑的雕刻线条流畅而优雅，刀工细腻而精美，为这座古老的宅院增添了别样的古朴雅致之美。石屏会馆仿佛是一幅精心绘制的画卷，将滇南建筑的独特韵味展现得淋漓尽致。每一处细节都散发着历史的沧桑和文化的底蕴，让人沉浸其中，感受着岁月的流转。站在这里，仿佛能够穿越时光，感受到古老时代的风貌和人们的生活气息。

【 **Historical background** 】

In the 10th year of the Republic of China (1921), Shiping people from all circles in Kunming were initiated by Mr. Yuan Jiagu and Mr. Zhang Zhijiang to provide temporary apartments and places for business meetings for Shiping students and business travelers studying in Kunming. The scale of the first expansion is very large, rebuilt into a combination of Chinese and western large-scale gate, stone carved semicircular gate, embedded with the "Shiping Guild Hall" banner, inside is the three-Jin Courtyard two-storey civil structure building, decorated simply. The front platform and steps are inlaid with bluestone.

After 2000, in order to protect the excellent historical building and restore the traditional style, 54 households living in the building moved out one after another. After the second restoration in April 2004, the Shiping Guild Hall recreatecl the glorious scenery of the past, and the old house still retains its ancient charm, and the heavy cultural heritage and long historical connotation are vividly reflected.

【历史背景】

民国十年（公元1921年）由袁嘉谷先生和张芷江先生发动在昆的石屏各界人士为在昆明读书的石屏籍学生和商旅人士提供临时寓所及商贸集会的场所，扩建了这所宅院。第一次扩建的规模甚为宏大，改建为中西合璧的大型门楼，石雕半圆形门坊上，嵌有“石屏会馆”横额，内是三进院二层土木结构楼房，装饰简洁。门前平台、踏步青石镶砌。

2000年后，为了保护优秀历史建筑，恢复传统风貌，居住在建筑中的54户居民陆续迁出。2004年4月第二次修复后的石屏会馆再现了昔日的盛景，修旧如旧的老宅古韵犹存，厚重的文化底蕴和悠长的历史内涵得到淋漓尽致的体现。这次修缮保留始建风貌，门、窗均为原物原状，墙墩经过加固校正后保留原貌。

石屏会馆完整地保留了它独特的历史风貌，是昆明老建筑修复工程中具有代表性的典范。

3.5 Yunnan Agricultural Exhibition Hall (Yunnan Science and Technology Museum) 5

In 1958, a new Yunnan Agricultural Exhibition Hall was built on the former site of the playground outside the Yunnan Army Training Hall. It was renamed Yunnan Provincial Exhibition Hall in 1981 and Yunnan Provincial Science and Technology Museum in 1984. It covers an area of 31,000 square meters, and the building area is 10,000 square meters. The main building is a two-storey arc reinforced concrete structure with Soviet-style architecture. There are 19 exhibition halls, with an exhibition area of 8,000 square meters. It was the largest science and technology exhibition and information exchange center with the best conditions in Yunnan Province at that time.

5

3.5 云南省农业展览馆（云南省科技馆）5

1958年在云南陆军讲武堂外操场的旧址新建云南省农业展览馆。1981年改为云南省展览馆，1984年改为云南省科技馆。占地面积31000平方米。建筑面积10000平方米。主楼为两层弧形钢筋混凝土结构，具有苏式建筑风格。内有展厅19个，展区面积8000平方米。是当时云南省内规模最大、条件最好的科技展览馆和信息交流中心。

建筑呈中轴对称式的“Y”字形布局，主楼庄重高耸，门廊共设六对圆柱，气势宏伟。不仅如此，在柱头和柱脚还设计了祥云和卷草花的纹饰，更显奢华。整座建筑弧形回廊舒缓而又优雅，回廊采用三段式设计，檐部、勒脚装饰浮雕，墙身简约大方。

云南省农业展览馆（云南省科技馆）是现存的昆明十大公共建筑之一。

3.6 Dongyuan Villa 7

Dongyuan Villa (No.7-9 Huashan East Road) is located on the east side of Wuhua Mountain and was built during the Republic of China period. Built on the hillside of Wuhua Mountain, Dongyuan Villa is against the mountain and has a terrace on the third floor. The building is of symmetrical design, both ends of the semi-hexagonal shape, brick and wood structure, blue brick tiles, thin lattice window.

In 1943, Song Jiajin invited a French architect to design and build this villa, which has a certain historical and artistic value. Song Jiajin, graduated from the Whampoa Military Academy, held an important position in the Northern Expeditions Army, and served as county governors of Baoshan County, Shiping County and Jinning County.

3.6 东苑别墅 7

东苑别墅（华山东路7—9号）位于五华山东侧，建于民国时期。五华山山坡上的东苑别墅，依山就势修建，其三层设有露台。建筑呈对称式设计，两端为半六边形造型，砖木结构，青砖碧瓦，细格条窗。

1943年，宋嘉晋邀请法国建筑师设计这座别墅，具有一定的历史、艺术价值。宋嘉晋，毕业于黄埔军校，在北伐军中任要职，曾先后担任过保山县、石屏县和晋宁县县长。

3.7 Yang Family Mansion, Huanghe Lane 6

Yang Family Mansion is located at No. 37 Huanghe Lane, East Huashan Road. Built in 1932, it was the private residence of Yang Ruxuan, a general of the Yunnan Army, and designed by a French architect.

The residence is a two-storey brick and wood structure of French architecture. The symmetrical design, the reliefs and lines of the building window covers and windowsills, the railings and columns of the terrace, the vases, and the multi-level stone steps at the entrance give the building a romantic and grand atmosphere. The roof is steep and high, and the windows are tall and evenly proportioned to the walls.

3.7 黄河巷杨氏公馆 6

杨氏公馆位于华山东路黄河巷37号，建于1932年，是滇军将领杨如轩的私人住所，由法国建筑师设计。杨氏公馆为二层砖木结构的法式建筑。采用对称式设计，建筑窗套、窗台的浮雕与线条，露台栏杆立柱、花瓶，入口的多级石台阶，赋予了建筑浪漫而宏伟的氛围。屋顶陡峭而高耸，窗户高大与墙面的比例均衡，使得建筑具有独特的魅力。

杨如轩（1895—1979年），字夷斋，云南宾川人。云南陆军讲武堂毕业，参加辛亥昆明重九起义、护国战争等。1927年“八一南昌起义”时，杨如轩任国民党第九军二十七师师长驻防江西临川，应朱德要求“临川让路”。1930年后，杨如轩曾任云南宪兵司令、云南防空司令部司令等职。新中国成立后，曾任云南省政协委员、云南文史馆馆员。

3.8 He Family Mansion, Dafuchun Street 8

He Family House is the house of the general He Shixiong of the Yunnan Army, located in 83 Dafuchun Street, built in the eighth year of the Republic of China (1919).

The courtyard is a typical traditional building in Dali, Yunnan Province, in the style of "three squares and one view wall" and "walking horse and turning-corner tower". The courtyard faces south, the main room and the two side rooms are connected by the corridor, the partition doors and windows, the square, the beams and columns, the eaves carved fine. The house is built with superb skills and beautiful environment: the white wall is ancient and elegant; the silk seam wall is polished with brick; the wisteria, the rockery, and the fish in the well pond.

3.8 大富春街何氏宅院 8

何氏宅院（将军楼）是滇军将领何世雄的宅院，位于大富春街83号，建于民国八年（1919年）。宅院为典型的云南大理"三坊一照壁""走马转角楼式"样式的传统建筑。院落坐北朝南，正屋与两厢房皆由回廊相连，隔扇门窗、枋额、梁柱、檐头雕刻精细。宅院建造技艺高超，环境优美，粉墙黛瓦古朴雅致，丝缝墙磨砖对缝，紫藤假山，水井池鱼。

该宅院为原滇军将领何世雄修建的私人住宅，旅缅侨商舒子烈于1945年收购此宅，后几易其主。何世雄，字子侯，云南云龙人，毕业于云南讲武堂，1919年任近卫军十四团团长，1924年任近卫军第三旅旅长，1927年重任第四混成旅少将旅长。

7

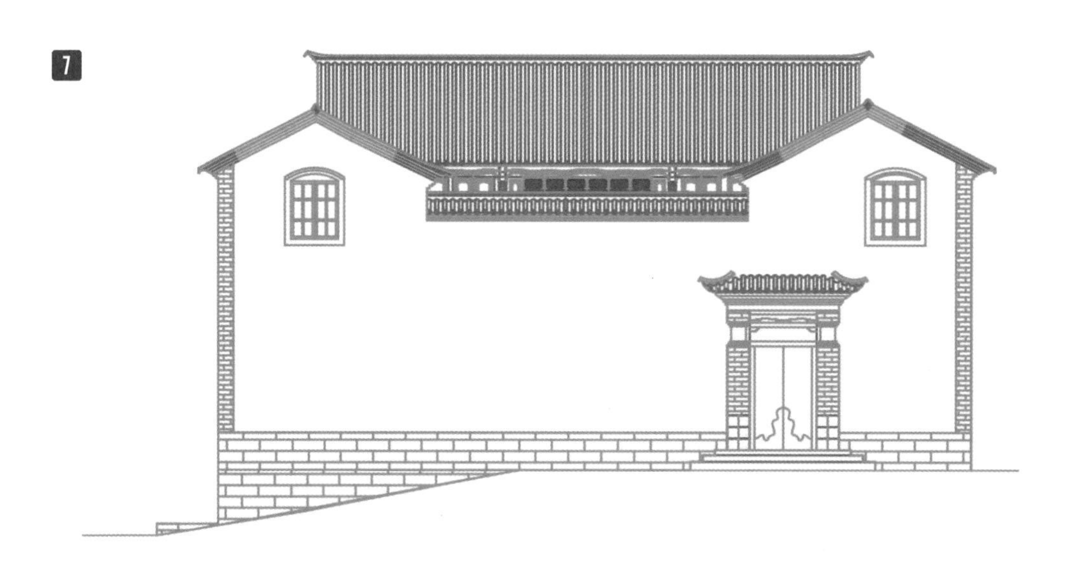

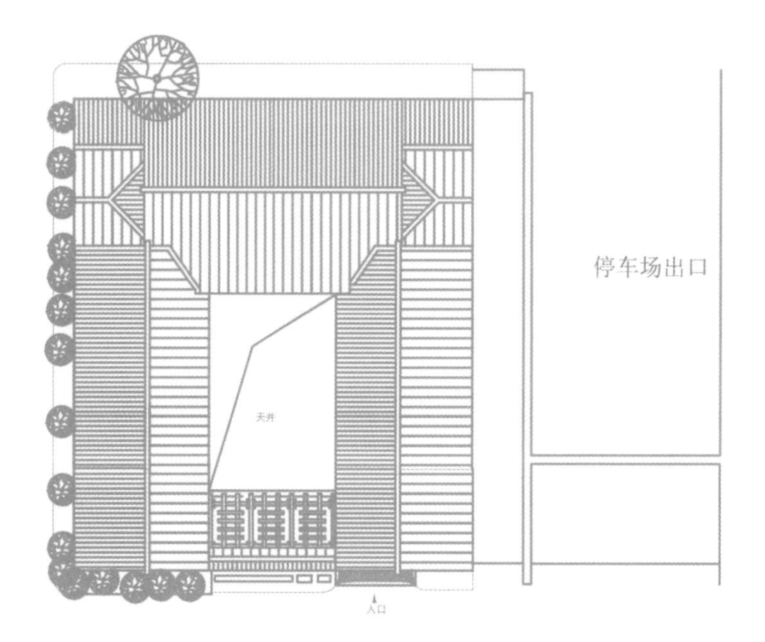

3.9 No. 3, Fuchun Street 9

Located in No. 3 Fuchun Street, built during the Republic of China period, the house is a traditional building with a brick and wood structure of “three squares and one view wall”. The carving techniques of the house eaves, railings, doors and windows, and inner corridors are fine and unique. Small wooden works such as columns, tiework, eaves and balustrades are also decorated with paintings. The lattice door is embedded with oval glass, and the stone drum pillar foundation is exquisite and exquisite. The style of the door square combines Chinese and Western elements, making the building shining.

3.9 富春街 3 号宅院 9

位于富春街的3号宅院，建于民国时期。宅院为“三坊一照壁”式砖木结构、坐北向南的传统建筑。宅院檐板、栏杆、门窗、内廊的雕刻工艺精细、别致。柱子、额枋、檐板、栏杆等小木作也均有彩绘装饰。格扇门上嵌有椭圆玻璃，石鼓柱础精致玲珑，门坊风格中西合璧，使得建筑熠熠生辉。

3.10 Nuclear industry business building 10

Yunnan Nuclear Industry Business Building, located at 182 West Dongfeng Road, was once Kunming Guest House of Yunnan Nuclear Industry 209th Geological Brigade. The building has a brick-concrete structure and Soviet style. The building has five floors and stands along the street. The interior has a modern elevator. After careful maintenance, the building is well preserved and exudes the charm of history.

3.10 核工业商务楼 10

位于东风西路182号的云南核工业商务楼，曾是云南省核工业二〇九地质大队昆明招待所。这座建筑采用砖混结构和苏式风格。建筑共有五层，沿街而立，内部设有电梯。经过精心维护，建筑保存完好，散发着历史的韵味。

4 东寺街

<片区>

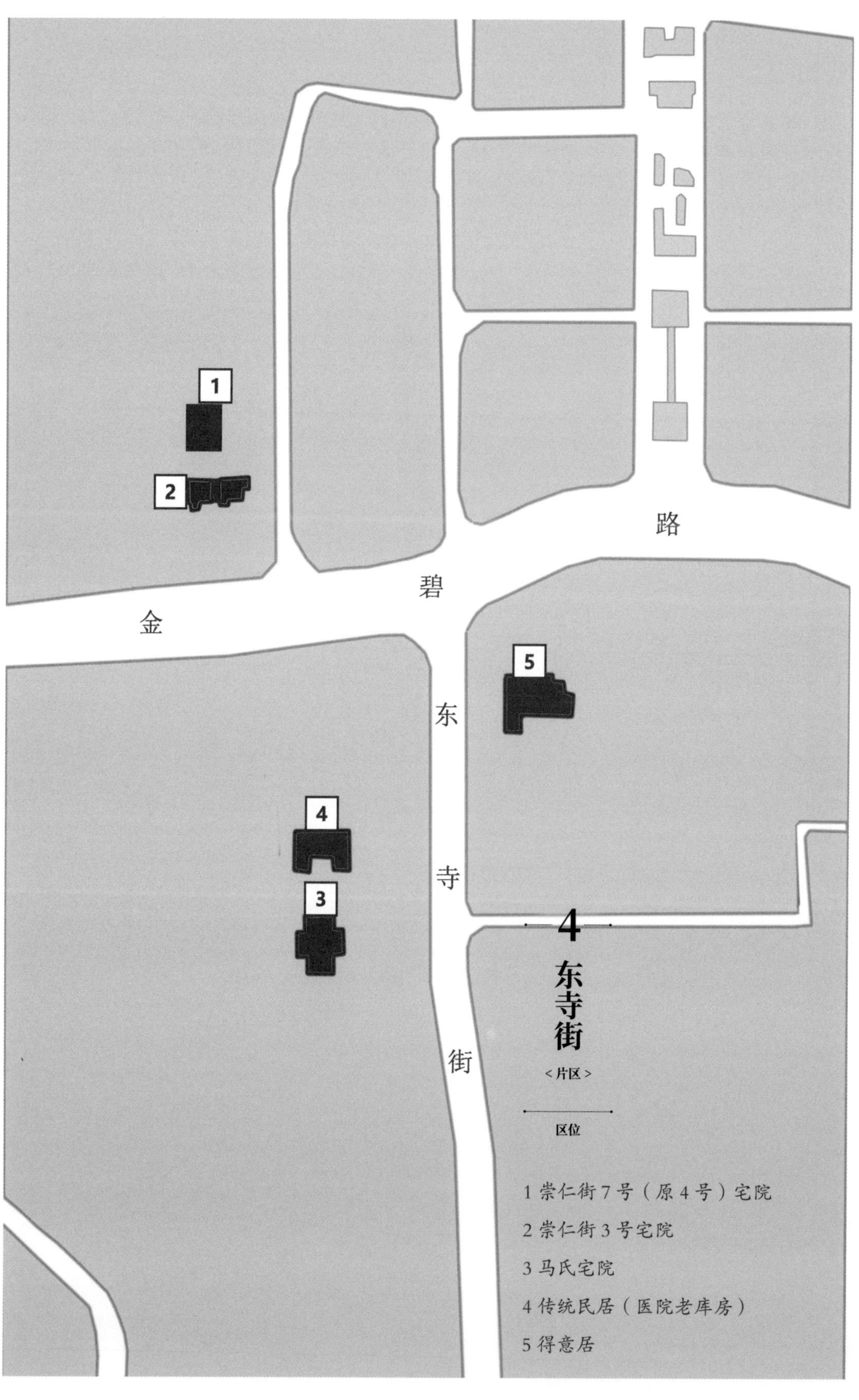
路
碧
金
东
寺
街
1
2
3
4
5
4
东寺街
<片区>
区位
1 崇仁街7号（原4号）宅院
2 崇仁街3号宅院
3 马氏宅院
4 传统民居（医院老库房）
5 得意居

4 东寺街片区

◈ 地名由来

东寺街

东寺街是一套古老的街道，形成于明清时期，得名于原东寺，即唐代的觉照寺（已毁），目前只剩下东寺塔。

崇仁街

这条街道的南起点是金碧路，北至顺城街。在清代初年，府城总盐店在这里开设，因此得名为盐店街。到了民国初年，改为现在的名字。

4.1 Ma’s House 3

Ma’s House is located at No. 5 Huajiao Lane, Dongsi Street. It was built in the ninth year of the Republic of China (1920) as the former residence of Mr. Ma Cong, a patriotic democrat. Facing south, this house is composed of front hall, main room, east and west chamers and wing. It shows the typical characteristics of Yunnan folk houses and adopts the layout of “four and five courtyards”. The small wooden works such as flower square, flower board, transparent carving and hanging, doors and wndows in the courtyard are beautifully carved, and the interior floor still retains a small part of the original French imported tiles, showing the exquisite art. The cast iron flower railing on the upper floor is engraved with the national flag pattern after the successful revolution of 1911, highlighting the spirit of the times.

4.1 马氏宅院 3

马氏宅院位于东寺街花椒巷5号，建于民国九年（1920年），为爱国民主人士马骢先生的旧居。这座宅院坐北朝南，由前厅、正房、东西厢房及耳房组成，展现了典型的云南民居特色，采用了“四合五天井”的布局。宅院内的花枋、花板、透雕挂落、门窗等小木作品都雕刻精美，室内地面还保留有小部分原法国进口花砖，展示了精湛艺术的工艺。楼上的铸铁花栏杆上刻有辛亥革命成功后的国旗图案，彰显了时代的精神。

【 **Historical background** 】

Ma Cong, Hui nationality, army Lieutenant general, Yunnan historical celebrity, had participated in the Revolution of 1911, the National Defense War, the Protection of the Law movement, the War of Resistance against Japanese Aggression, etc. He was one of the 39 senior generals of the Yunnan Army when he first defended the country. In the defense of the law he led his troops to Sichuan, and the warlord decisive battle. During the Anti-Japanese War, he was deputy commander of the Yunnan Military Command and organized the Yunnan Resistance and the Yunnan-Burma Resistance. At the same time, he actively participated in the democratic revolution and welcomed the birth of New China. He has held important posts, including Chief of the General staff, acting governor, salt transport minister, director of Industry department, Director of Finance Department, deputy commander of the military control area, etc. He also served as president of the Yunnan Muslim Salvation Association. After the founding of New China, he served as a member of the Civil Affairs Committee of the Southwest Military and Political Committee.

【历史背景】

马骢，回族，陆军中将，曾参加辛亥革命、护国战争、护法运动、抗日战争等。护国首义时是39位滇军高级将领之一；护法时率部出师四川，与军阀决战。抗战时任云南军管区中将副司令，组织出滇抗战和滇缅抗战。同时积极参与民主革命，迎接新中国的诞生。历任要职，包括滇军总参谋长、代理省长、盐运使、实业厅厅长、财政厅厅长、军管区中将副司令等。还担任云南回教救国协会会长。中华人民共和国成立后，担任西南军政委员会民委委员、云南省政协常委、省民委副主任等。

4.2 Traditional Residence (Hospital old warehouse) 4

This traditional residence is located in the First People's Hospital of Yunnan Province on Jinbi Road, the former site of the "Jinbi Park" built in the late Qing Dynasty. In 1939, the Yunnan Provincial Government started the Hospital project, during which the building was also constructed. However, due to another expansion of the hospital in 2009, the building was completely relocated.

This residence adopts the traditional "three squares and one view wall" civil structure, showing the typical two-storey residential style. The spacious patio brings ventilation and natural light throughout the building. Small wooden carvings such as cornice, columns and corridors are simple and exquisite.

4.2 传统民居（医院老库房）4

这座传统民居坐落于金碧路云南省第一人民医院内，原为清末兴建的“金碧公园”旧址。在1939年云南省政府兴建医院时，一并修建了此处院落。然而，由于2009年医院再次扩建，该建筑整体迁建。这座民居采用传统的“三坊一照壁”土木结构，展示了典型的二层民居风格。宽敞的天井为整个建筑带来了通风和自然光线。屋内的檐口、柱子和廊道等小木作雕刻简洁而精美。

4.3 得意居 5

得意居位于昆明金碧商城内，最初建于清末，经过1998年的修缮，整座建筑保存完好。得意居是一座典型的“三坊一照壁”式建筑，正房三层和厢房两层。整座建筑坐北朝南，以其古朴典雅的外观和精美的细节吸引着人们。起翘的屋顶，屋檐下“山水烟云”的彩绘，大门须弥座雕饰古朴，与精美的石雕花台、鱼池、柱础以及院墙上的影壁“福”“寿”相得益彰。砖雕小品玲珑别致。院内的雕梁画栋、华丽装饰，门窗、挂落、额枋、梁头上刻有栩栩如生的凤穿牡丹、兔含香草、鹤鹿同春等图案，令人叹为观止。整体建筑融合了中西建筑的精华，成为昆明传统民居中的杰作。

4.4 No. 3 Chongren Street 2

The two villas at No. 3 Chongren Street, built during the Republic of China, were originally the American Consulate in Kunming during the Anti-Japanese War, and are relatively intact. Chongren Street No. 3 has two modern Western-style villas, using a layout of one house. The villas have three storeys of brick-concrete construction, the exterior walls are covered with washed stone materials, and the interior is ornate and detailed. The long balconies on the second floor and the semicircular balconies on the third floor are arranged in a staggered manner, making the building facade unique and varied. The exterior walls of the buildings are washed stone and bean sand stone, showing an elegant gray tone, and brown lines, adding to the richness and beauty of the exterior walls.

4.4 崇仁街 3 号宅院 2

崇仁街3号的两处别墅，建于民国时期，原为抗战时期美国驻昆领事馆，保存较完整。崇仁街3号拥有两栋西式近现代风格的别墅，采用了一进两院的布局。这些别墅共有三层，采用砖混结构，外墙为水洗石材料，内部装饰华丽而精细。二楼的长条形阳台和三楼的半圆形阳台错落有致地交错排列，使建筑立面独特而富有变化。建筑外墙采用水洗石和豆砂石，呈现出优雅的灰白色调，还有棕色线条装饰，增添了建筑外墙的丰富和美观。

4.5 No. 7 Chongren Street 1

The house is a typical traditional residence with four and five patios, a two-storey civil structure and a hard tiled roof. There are three rooms in the main room, five rooms in the secret room, three rooms on each side of the east and west wings, and five rooms in the front room. Bluestone floors are laid outside, and stone is also used for the bases of the steps and columns. The facade of the building uses Western curved roof windows, the door is designed in European style, the top of the stone columns are arched, and Roman columns and semicircular arch doorways are added. The entrance of the building and the windows of the facade have elements of Western architecture.

Built in 1920, No. 7 Chongren Street originally belonged to the private residence of furrier Zhang Ziyu. Between 1922 and 1928, the American Consulate in Kunming rented the building, which was later purchased by the Ma Chaoqun brothers of the Chuxiong Mercantile Gang. It served as the office of the Wuhua District Armed Forces Department and Chongren Police Station.

4.5 崇仁街 7 号（原 4 号）宅院 1

这座宅院是一座典型的传统民居，四合五天井，走马转角楼，土木结构两层，硬山瓦屋顶。正房明间有三间，暗间有五间，东西两侧耳房各面阔三间，前厅则面阔五间。室外地面铺设青石地面，台阶和柱子的基座也采用了石材。建筑的外立面采用了西式弧顶窗，大门则采用欧式设计，石柱顶部有拱形装饰，还加上了罗马柱和半圆拱顶门套。建筑的入口和外立面的窗户具有西方建筑的元素。

崇仁街7号建于1920年，最初属于皮货商人张子玉的私人住宅。在1922年至1928年期间，美国驻昆明领事馆租用了这座建筑，后来被楚雄商帮的马超群兄弟购买。中华人民共和国成立时，曾作为五华区武装部和崇仁派出所的办公场所。

5
昆明动物园

< 片区 >

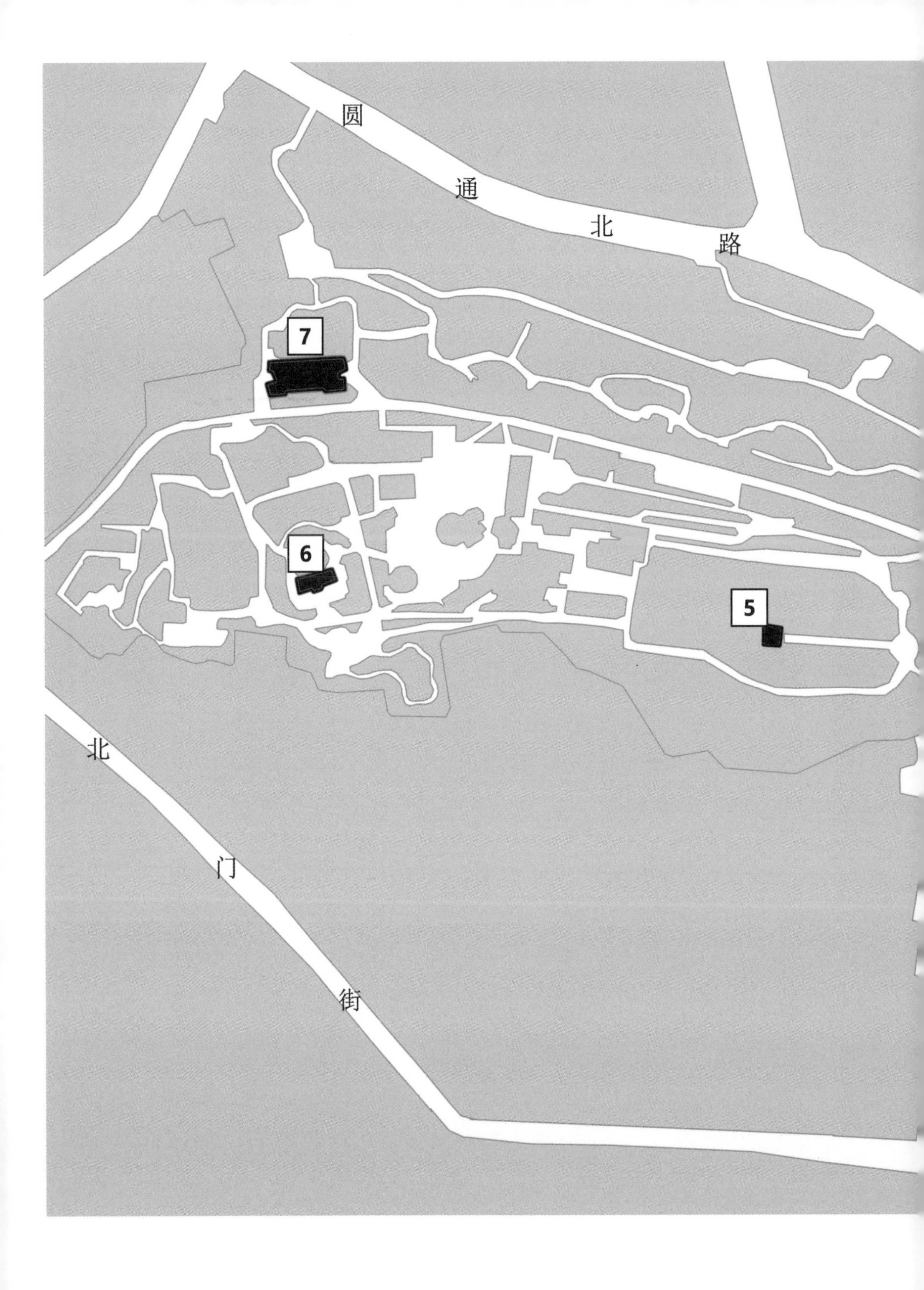
圆
通
北
路
7
6
5
北
门
街

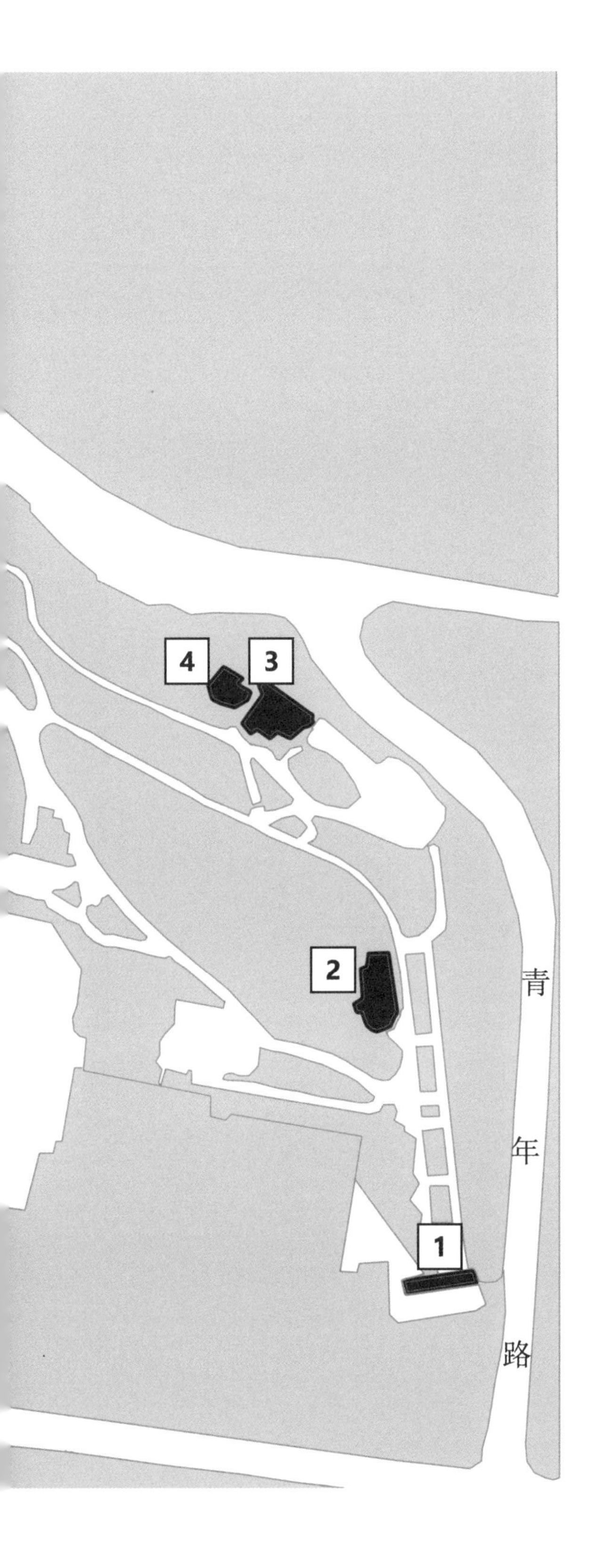

5
昆明动物园
<片区>

区位

1 昆明动物园大门

2 昆明动物园长臂猿馆

3 昆明动物园热带猴馆

4 昆明动物园金丝猴馆

5 缅甸战役中国阵亡将士纪念碑

6 唐家花园遗址

7 昆明动物园亚洲象馆

5 昆明动物园

◈ 昆明动物园的历史

动物园现址的历史最早可以追溯到元代。元大德五年（1301年）至延佑六年（1319年），云南行中书省左丞阿昔思，在螺峰山南麓创建圆通寺，由山麓沿盘坤岩峭壁凿盘崖磴道上山，山上建藏经阁。自圆通寺建成，螺峰山又称圆通山。

明洪武十五年（1382年），筑昆明砖城，将圆通山圈入城内。明成化年间重修圆通寺，将山巅之藏经阁改为接引殿。清康熙七年、八年（1668年、1669年），吴三桂重修圆通寺，将山门南移百步至街面，兴建胜境坊，建水池中弥勒殿，架两座石桥连通池心岛，重修山巅接引殿及三洞券门韦驮殿。

民国初年，唐继尧在螺峰山西麓建梅园别墅，称“红楼”。1927年5月唐继尧病故，1930年为其在梅园东面建墓，俗称“唐坟”。1927年，螺峰山改建为公园。以后增建四方亭、聂耳亭，除圆通寺大门外，增辟北门街、北仓坡入口。1937年抗日战争爆发，圆通山成为国民党云南地方防备司令部、防空司令部，成为军事基地，园林荒芜。1946年1

月，军队撤出圆通山，当时山上有四方亭、三角亭、雨苍亭、月石草亭；四方亭南面留有一片抗日阵亡将士陵墓，东面耸立有石砌围栏“滇西抗日阵亡将士纪念碑”。

1951年，中国人民解放军军管会将圆通山移交省博物馆筹备处管理。博物馆筹备处设在圆通寺内。1953年，圆通山建动物园。7月，大观楼饲养的野生动物正式移交动物园，计6种7只（即黑鹿、狗熊、猴子、猫头鹰、狐狸各 1只，孔雀2只）。1954年元旦节，动物园正式开放。

1956年市建设局投资2.3万元建动物笼舍。1957年昆明东城墙拆除，动物园新建东大门，正对拆除城墙新建的青年路。1958年4月11日，省博物馆将圆通寺移交圆通山管理。1959年初正式命名为圆通动物园。1986年更名为昆明动物园。1958年至1965年，逐年增添动物笼舍。仅1961年就投资100万元，建大型动物笼舍，1962年竣工。

5.1 Gate of Kunming Zoo 1

The main gate of Kunming Zoo is a reflection of the influence of former Soviet architecture on Kunming's urban construction in the 1970s: Witness the development of Kunming's special animal park. It imitates of the former Soviet Union's architectural style, light yellow color, left and right axis symmetry, plane rules, four columns and two rows of columns to form three rooms, east and west sides of the two supporting rooms. There is a "three-stage" structure, including the eaves, the wall body, and the leg of the three parts.

1

5.1 昆明动物园大门 1

昆明动物园大门是20世纪70年代昆明城市建设中受苏联建筑风格影响的体现：见证昆明城市动物专类公园的发展。苏联建筑风格，淡黄色为主色调，左右呈中轴对称，平面规矩，四列两行柱子形成三开间，东西两侧分别有两间配套用房。有“三段式”结构，包括檐部、墙身、勒脚三个部分。

5.2 Gibbon Hall of Kunming Zoo 2

Gibbon Hall of Kunming Zoo is a witness to the history of the construction and development of Kunming special animal park. In 1953, it began to prepare for the construction of a zoo. Kunming Zoo and its animal cage buildings are the material carriers of the precious memories of old Kunming people, reflecting the characteristics of domestic architecture in the 1980s. The plane is a combination of semi-circular and regular rectangle, and closely matched with the terrain, retaining the rockery, and the overall height is scattered. The interior atrium is a triple-height space surrounded by external corridors as viewing spaces. The corridor, entrance, and flower shelves are combined with external plants, which have certain characteristics of garden architecture.

2

3

5.2 昆明动物园长臂猿馆 2

昆明动物园长臂猿馆见证了昆明动物专类公园的建设发展历史。1953年昆明开始筹建动物园，昆明动物园及其中的动物笼舍建筑是老昆明人珍贵记忆的物质载体，反映20世纪80年代我国建筑风格特征。平面为半圆形与规则矩形结合，并与地形紧密配合，保留假山石，整体高低错落。内部中庭三层通高空间，外部走廊环绕作为观赏空间。走廊、入口、花架与外部植物结合，具有一定园林建筑特征。

5.3 昆明动物园热带猴馆 3

昆明动物园热带猴馆平面结合地形为圆形与半圆形组合，内部圆形为服务用房，南侧外围半圆形建筑围绕，作为动物活动用房立面为玻璃。整体结合地形错落有致，满足观光、后勤服务与地形的需求。

5.4 Golden Monkey House of Kunming Zoo 4

Golden Monkey House of Kunming Zoo is a witness to the construction and development history of Kunming special animal park. In 1953, Kunming Zoo and its animal cage buildings are material carriers of the precious memories of old Kunming people, reflecting the architectural characteristics of the early 21st century. The plane is three rectangles successively combined, the whole is divided into north and south parts: the south is used as the animal activity room, the facade is glass; the north is the logistics service room. It is combined with the terrain, scattered to meet the needs of sightseeing, logistics services and terrain.

5.4 昆明动物园金丝猴馆 4

昆明动物园金丝猴馆平面为三个矩形依次递进组合，整体分南北两部分，南部作为动物活动用房，立面为玻璃，北部为后勤服务用房。结合地形，错落有致，满足观光、后勤服务和地形的需求。

5.5 昆明动物园亚洲象馆 7

昆明动物园亚洲象馆一字型平面，东西两侧基本对称，中部高两侧低。北侧为动物活动用房，南侧为后勤用房，有东西两个入口。一层挑檐，勒脚，石材镂空隔断和大窗格均保留，立面材料为水刷石和不规则石材贴面。整体古朴素雅。

5.6 唐家花园遗址 6

云南护国运动领导人唐继尧将军居所“唐家花园”的珍贵遗址。民国时期修建，花园外围建有会泽亭、走廊、厅堂，中央通过石桥、水池假山等营造园林景观，体现一定的园林艺术与名人遗风。

5.7 Memorial to the Chinese Soldiers Killed in the Burma Campaign 5

The Memorial to the Chinese Soldiers Killed in the Myanmar Campaign is the physical witness of the history of the Anti-Japanese War in west Yunnan. Also known as the "An LAN Memorial Tower", The Memorial is located at the highest point of the Yuantong Mountain to the north of the Luofeng Pavilion in October 4, 1944, completed in February 1945. The stela was destroyed in the 1950s and rebuilt on its original site in 2013. Built of white stone, the lower base is 18 meters high, the three-sided column body; the upper tablet body is 8.54 meters high, triangular prism. The base is inlaid with three stone tablets, and the front is engraved with the bust of General Dai Anlan, and the inscriptions of Xu Tingyao, Chiang Kai-she, He Yingqin, Wei Lihuang, and Long Yun. The other two tablets are inscribed with inscriptions written by Lieutenant General Du Yuming, commander of the defense of Kunming, detailing the harsh natural environment of the Burma campaign, the fierce battle, the national integrity of the soldiers, and the sacrifice of Dai Anlan and other generals and soldiers.

5.7 缅甸战役中国阵亡将士纪念碑 5

缅甸战役中国阵亡将士纪念碑是滇西抗战历史的实物见证。缅甸战役中国阵亡将士纪念碑，又称“安澜纪念塔”，位于圆通山最高处螺峰阁北面，于1944年10月4日奠基，1945年2月落成。石碑于20世纪50年代被拆毁，2013年原址重建。白石砌成，下部基座高18米，三边柱体，上部碑体高8.54米，呈三角菱形。基座镶嵌三块石碑，正面镌刻戴安澜将军半身遗像，徐庭瑶、蒋介石、何应钦、卫立煌、龙云的题词。其余两块石碑镌刻陆军中将、昆明防守司令杜聿明所撰碑文，详细记述缅甸战役恶劣的自然环境、惨烈的作战经过、将士的民族气节，及戴安澜等将士牺牲的情况。

50

6

东风路

<片区>

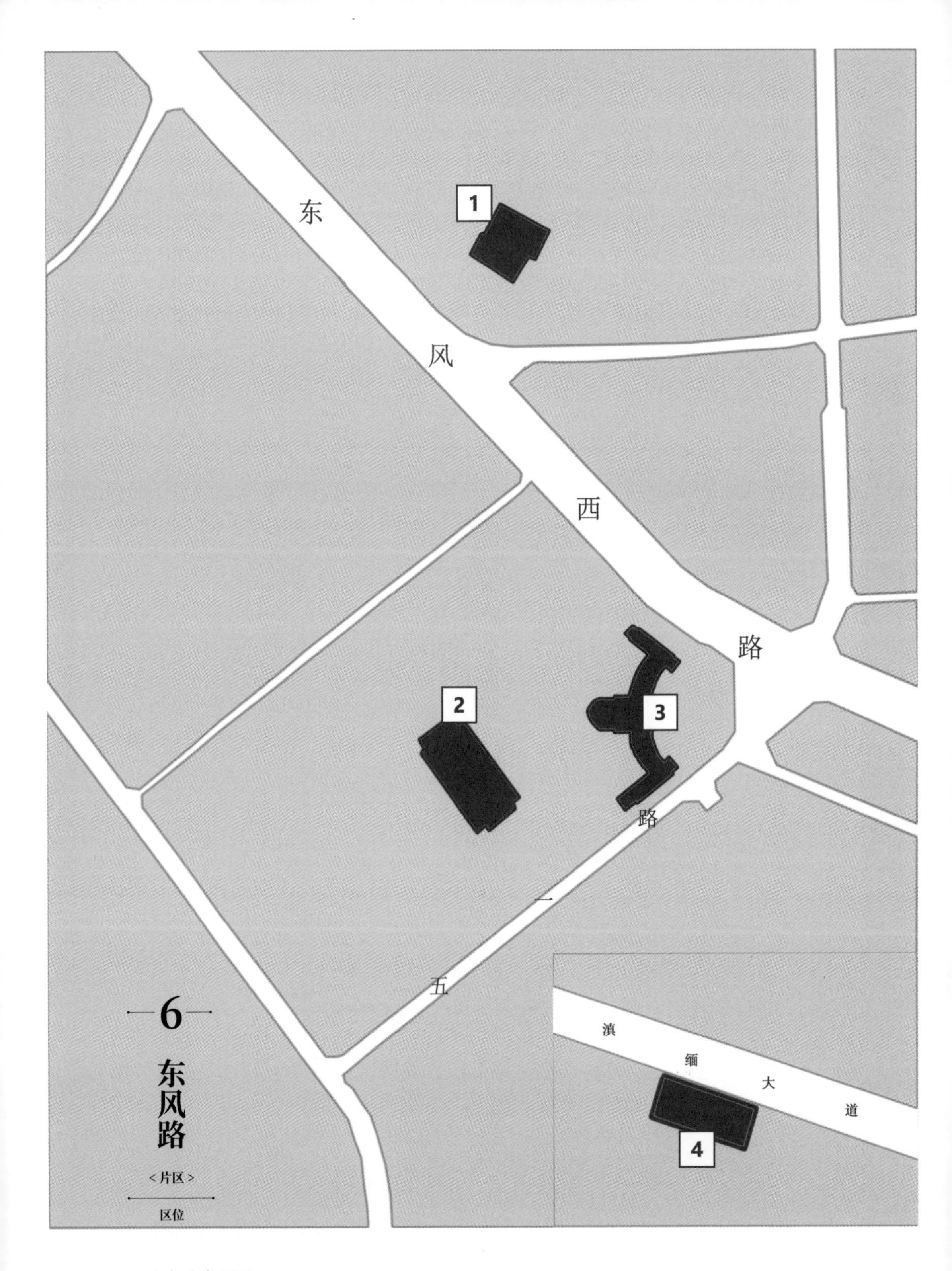

1 云南艺术剧院

2 国防剧院

3 云南省博物馆

4 西南联大教学楼和援华美军招待所旧址

6 东风路片区

◈ 地名由来

东风路，东起环城东路，西至南屏街与护国路交会。1952—1956年分段建成。护国路口至北京路一段曾取名“南太路”。东风西路，东起近日公园，与南屏街相望，北抵龙翔街东口，中与五一路、大观街、昆师路、新闻路、人民西路、翠湖南路等街路相通。明清时为南门至大西门、小西门间的城垣和护城河。1952年拆除城墙，填平河道，筑成市区交通主干道之一，取名近西路。

1960年以“东风压倒西风”之意，改称东风路，1980年以路处市区东面、西面之故改为今名。

◈ 昆明十大公共建筑

昆明十大公共建筑是指为庆祝中华人共和国成立十周年，20世纪50年代建成的苏式建筑，也被誉为昆明的“十大建筑”。那个年代，中国、苏联两国友好相处、亲密往来，这些建筑具有明显的苏式建筑风格，散发出时代的气息。

昆明十大建筑，主要分布在东风路沿线和翠湖的周边。10座建筑分别为：云南省博物馆、云南省农业展览馆（云南省科技馆）、云南省艺术剧院、昆明邮电大楼、昆明饭店老楼、云南省体育馆大楼、东风大楼、云南饭店（老楼）、昆明百货大楼（老楼）、翠湖宾馆（老楼）。

云南省科技馆、云南省艺术剧院、云南省博物馆、昆明邮电大楼、昆明饭店老楼、云南省体育馆大楼，至今这6座建筑都保存完整，已被列入文物保护单位或历史建筑。其余4座建筑已拆除，或在原址重新建设。

6.1 Yunnan Provincial Museum 3

The old museum of Yunnan Province is located on the west side of the intersection of Wuyi Road and West Dongfeng Road in central Kunming. Founded in 1951 and started in 1958, it was originally intended to be the Kunming Military Museum, which was later named the Yunnan Provincial Museum in 1959 and completed in 1964, making it one of the first national first-class museums. The building has a distinct Soviet-style architecture and is regarded as the first of the top ten buildings in Kunming. The provincial museum has a total of nine floors (six above ground, three underground floors, two wings of three floors). It is an are building of a pyramid-style top, with reinforced concrete structure and the total height of 65 meters. The top of the building is designed with a red flag as the base and a golden five-star spire at the top.

6.1 云南省博物馆 3

云南省博物馆的老馆位于昆明中部的五一路和东风西路交叉路口西侧。1951年筹建，1958年动工，原拟建成昆明军事博物馆，后在1959年定名为云南省博物馆，1964年建成，是首批国家一级博物馆。该建筑具有明显的苏联式建筑风格，被视为昆明十大公共建筑建筑之首。博物馆共九层（地上六层、地下三层、两翼三层），攒尖塔式弧形建筑。钢筋混凝土结构，建筑总高65米。建筑顶端设计了以红旗为基座、顶部为金色五星的尖塔。建筑主体呈弧形布局，前廊排列12根水洗石圆柱。建筑典雅，庄重大方，气势雄伟。主体建筑面积6400平方米，展厅面积2400平方米。在当时是云南省最大的综合性博物馆。2014年，云南省博物馆搬迁至广福路的新馆，老馆场地也被移交云南省美术馆。

6.2 Yunnan Art Theater 1

Yunnan Art Theater is located at 138 West Dongfeng Road. It was built in 1957 with Soviet-style architecture. It covered an area of about 9,500 square meters, the construction area of 3,803 square meters, with 1200 seats, two floors of rectangular office rooms on both sides. It connected by corridors, symmetrical form the east and west wings, the front of the central hall set 4 washed stone colonnades, hollow sculpture peacock tea flower pattern between the columns under the eaves.

Around the 1930s, the Operatic Circle group set up a stage here, singing Beijing Opera and Yunnan opera, so it was called "group stage". The theater covered an area of 2,000 square meters and had 1,155 seats inside. In addition, there was a lounge on the west side of the theatre and a small concert hall upstairs in the foyer. At that time, it was one of the larger and better equipped theaters in Kunming.

6.2 云南艺术剧院 1

云南艺术剧院位于东风西路138号，建于1957年，苏式建筑风格。占地面积约9500平方米，建筑面积3803平方米，拥有1200个座位，两侧两层长方形办公用房，通廊连接，相互对称形成东西两翼，中央大厅前沿设4根水洗石柱廊，檐下柱间镂空雕塑孔雀茶花图案。云南艺术剧院是现存的昆明十大公共建筑之一。20世纪30年代前后，梨园班子在这里搭建了舞台，演唱着京剧和滇剧，因此被称为“群舞台”。而在1937年前后，为了纪念“重九”起义的名将蔡锷（字松坡），松坡中学在此地建立起来。到了1954年，这座建筑进行了改建，1957年竣工后，被命名为云南艺术剧院，然而在1966年后改名为红星剧院，直到1980年才恢复原名。整个剧院占地面积达到2000平方米，内部设有1155个座位。此外，剧院的西侧还设有休息室，前厅楼上还有一个小音乐厅。当时，它是昆明市内较大且设备较好的剧院之一。

6.3 国防剧院 2

国防剧院位于五一路68号，是新中国时期昆明典型公共建筑代表，具有明显的苏式建筑风格特征，保存现状良好，气势雄伟。

6.4 Former Site of Southwest Associated University Teaching Building and U. S. China Relief Military Guest House 4

The former site of the teaching building of Southwest Associated University and the Guest House for American Troops in China during Anti-Japanese War is located at No. 280 Yunnan-Burma Avenue. Built in 1936, it was originally the teaching building of Kunhua Agricultural School. In 1938, the College of Science, the College of Arts and the College of Law and Business of Southwest Associated University rented the teaching building and the east and west dormitories of Kunhua Agricultural School as the campus. With the outbreak of the Pacific War, the U.S. Military China theater Command set up a branch in Kunming. In December 1941, the National Government building was assigned to the Eleventh Guest House of the United States Army.

6.4 西南联大教学楼和援华美军招待所旧址 4

西南联大教学楼和援华美军招待所旧址位于滇缅大道280号，建于1936年，最早为昆华农校教学大楼。1938年西南联合大学的理学院、文学院和法商学院租用昆华农校教学大楼及东西寝室做为校舍。随着太平洋战争的爆发，美军中国战区司令部在昆明设立了分部。1941年12月，国民政府将校舍拨给美军第十一招待所使用。

昆华农校教学楼融合了中式传统屋顶和西式外立面，使其具有中西合璧的建筑风格。该建筑采用了重檐歇山顶作为屋顶，外立面的西式条窗、拱券式门厅墙、巴洛克式圆柱门廊等都充分展示了西式建筑的特色。在建筑材料的选择上也非常讲究，楼内的25级台阶全部采用了1米长的整石砌筑，外墙厚达60厘米的青石砌筑，外窗过梁也采用了整石铺设，建筑墙体基石上还雕刻着精美的檐边装饰。这些设计和材料的运用使得该建筑更加精致、耐久。

7

巡津街

<片区>

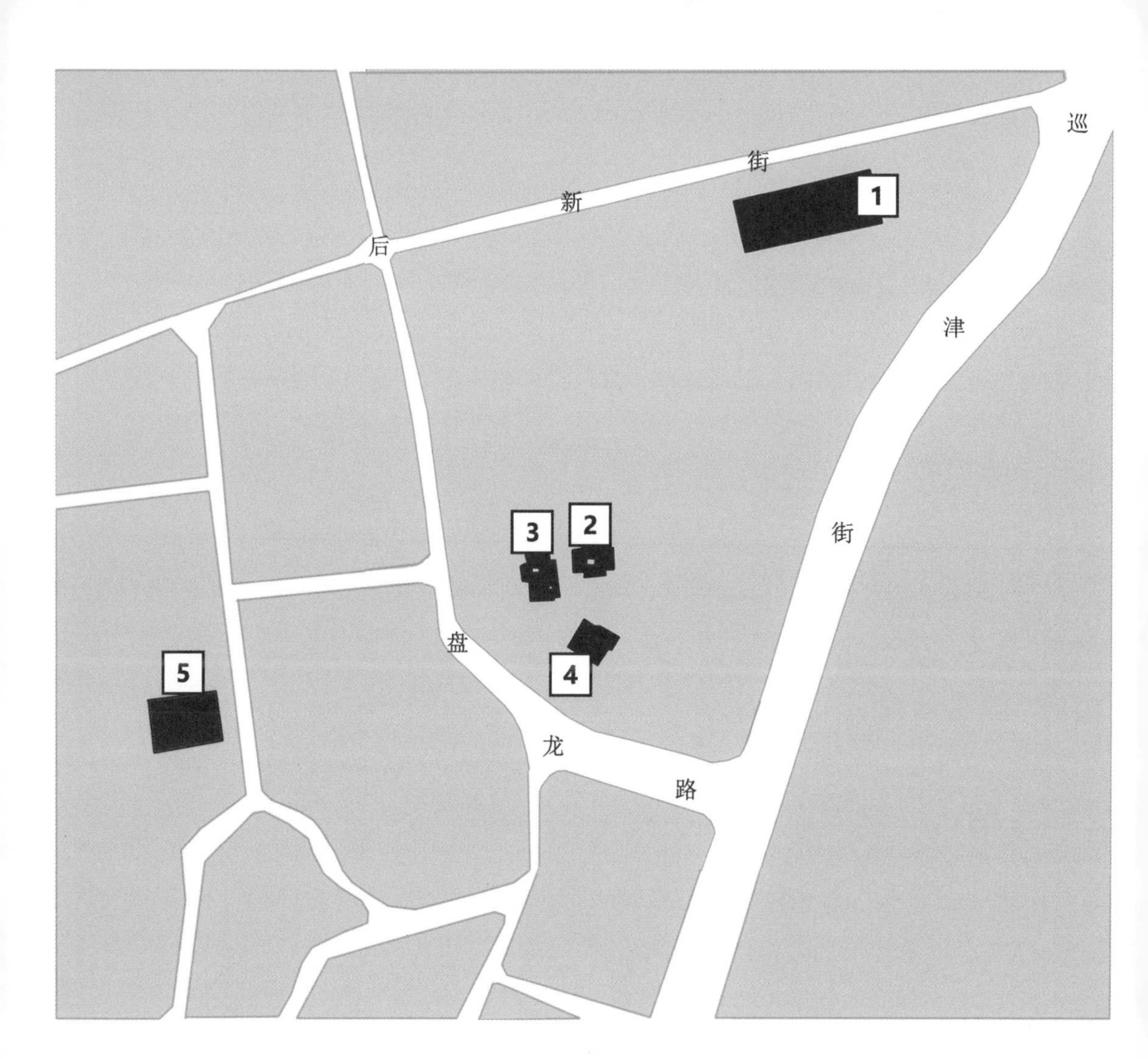

7

巡津街

<片区>

区位

1 甘美医院旧址

2 后新街 7 号别墅

3 后新街 8 号别墅

4 盘龙路 17 号别墅

5 盘龙阁

7 巡津街片区

◈ 地名由来

巡津街

巡津街位于盘龙江西岸，北起得胜桥，南至双龙桥。由于紧邻盘龙江，明代时被称为“大河埂”或“云津堤”，因为沿着河堤可以观赏到“云津夜市”。到了1920年，这条街道改称为现在的名字——“巡津街”。其中，“巡”一词意味着观察和预防，“津”则指的是盘龙江。

7.1 Former Site of Ganmei Hospital 1

Located on the west bank of Panlong River, Kunming, No. 35 Xunjin Street was once the site of Ganmei Hospital. Built in 1931, the general hospital was known as the "Noble Hospital" because it mainly served foreigners as well as Yunnan's dignitaries, military figures and wealthy classes. Ganmei Hospital is one of the important representative buildings for Western medicine to enter Kunming, and it is also one of the few French-style buildings in the Republic of China that have been preserved to the present day.

The Hospital, also known as the French Consular Hospital, was managed by the French Consulate, whose director was appointed by the French government. The medicine needed for the hospital was transported from Vietnam to Kunming by the French Consulate in Yunnan. Ganmei Hospital had internal medicine, surgery, gynecology and eye, ear, nose and throat and other specialties, and the equipment was relatively advanced. Wards were mainly concentrated on the first and second floors.

7.1 甘美医院旧址 1

位于昆明盘龙江西岸的巡津街35号，曾是甘美医院旧址。这座综合性医院于1931年建成，它因为主要为外国人以及云南的政要、军事人物和富有阶层提供服务，被称为“贵族医院”。甘美医院是西医进入昆明的重要代表性建筑之一，也是民国时期为数不多的保存至今的法式建筑之一。

建筑的主体有两层，局部还有三层。一层和二层的前后都有拱券式的外走廊，石栏杆上装饰着宝瓶柱。三层的双坡屋顶形成了三角形的山墙，与横向的二层形成了高低错落有致的对比。虽然现在的建筑已经将坡形屋顶改为了平顶，但整体结构基本上保留了原貌。

“甘美医院”还被称为“法国领事医院”，它由法国领事馆管理，院长是由法国政府委派的。医院所需的药品是由法国驻滇领事馆从越南运送到昆明。甘美医院设有内科、外科、妇科和眼耳鼻喉科等专科，设备相对较为先进。病房主要集中在一楼和二楼。

到了1950年8月，甘美医院被云南大学医学院接管，并更名为云南大学附属医院。随后在1958年，它再次更名为“昆明市第一人民医院”。

7.2 Panlong Pavilion 5

Panlong Pavilion, located in Xunjin Street, was built in the 1930s. It used to be the residence of Li Peitian, director of Finance Department and Civil Affairs Department of Yunnan Province during the Republic of China, and existed as the main building of private garden Jingyuan.

The Panlong Pavilion attracts people with its unique architectural style and exquisite details. It is a traditional two-storey building with a structure of brick, wood and stone. The broad side has eight rooms wide and three rooms deep. The exterior walls are made of square stone masonry, showing a solid and stable temperament.

7.2 盘龙阁 5

盘龙阁，坐落于巡津街，建于20世纪30年代。它曾是民国时期云南省财政厅厅长、民政厅厅长李培天的住宅，作为私人园林静园的主体建筑而存在。

盘龙阁以其独特的建筑风格和精美的细节而引人入胜。它坐西朝东，采用砖木石结构，是一座传统的二层建筑。宽阔的面阔有八间，进深有三间。外墙采用方形石块砌筑，展现出坚实而稳定的气质。

踏入环绕着石栏的七级台基，这里是作为盘龙阁门廊的雨亭迎接着人们的到来。雨亭顶部气派的歇山顶，雨亭下方则雕梁画栋、描金彩绘，展现出威严富贵的气息，却不失雅致之感。一步之间，仿佛置身于尊贵与优雅之境。

整座建筑以两翼对称的设计呈现，如同一座园林式建筑的精巧构思。建筑的两端头是半山亭，而人们可以在二楼的开敞半山亭中欣赏室外的美景。这种设计将自然与建筑巧妙地融合在一起，为人们带来愉悦的视觉享受。

外墙的门窗融入了西式建筑元素，二楼窗框呈半圆拱券样式，石线条从二楼延伸至一楼，形成宽大的窗套。这些细节不仅展示了盘龙阁的雄伟和精心设计，更凸显了它作为中国私家园林建筑的精巧之处，以及建筑本身所散发的奢华气息。

【Historical background】

Li Peitian (1895-1975), addressed respectfully as Zihou, brother of Li Peiyan in Binju Street, was an important figure in the modern history of Yunnan. After graduating from No. 1 High School of Yunnan Province, he went to Japan to study at the Political Economy Department of Waseda University. After returning to China, he taught at Beijing University of Political Science and Law, and then returned to Kunming in 1921, where he served as head of the Education Section of the Municipal Supervision Office, and also served as a teacher at the Provincial No. 1 Middle School and the private Chengde Middle School. In 1929, he was transferred to the office of the Yunnan Provincial Government in Nanjing and served as a member of the Mongolian-Xizang Committee, where he held many important positions. In 1936, Li Peitian was transferred back to Kunming, and successively served as director of the Yunnan County Head Training Institute, the director of the Civil Affairs Department, director of the Yunnan County Administrative Personnel Training Group, director of the Food Bureau, and director of the Financial Department until 1945, when the provincial government was reorganized.

【历史背景】

李培天（1895—1975年），字子厚，宾居街李培炎之弟，是云南近现代史上的重要人物。毕业于云南省立第一中学后，前往日本留学并就读于早稻田大学政治经济系。回国后，曾在北京政法大学任教，随后于1921年回到昆明，在市政督办公署任教育科长，并兼任省立一中和私立成德中学的教师。他于1929年调任云南省政府驻南京办事处处长，兼任蒙藏委员会委员，担任了多个重要职务。1936年李培天调回昆明，先后任云南县长训练所所长、民政厅厅长、云南省县行政人员训练团教育长、粮政局局长、财政厅厅长，直到1945年省政府改组后才去职。

李培天积极参与抗战和民主革命，为云南的抗日武装做出了重要贡献。1949年后，他参加了国民党革命委员会。

Villa at 7, Houxin Street; Villa at 8, Houxin Street; Villa 17, Panlong Road

In the south roadway of Kunming First People's Hospital, there are three villas of similar style under the shade of green trees, that is, No. 7 Houxin Street, No. 8 Houxin Street and No. 17 Panlong Road. These buildings were built in the Republic of China period, for the brick and wood structure of the two-storey building.

Villa7, Panlong Road was the private residence of the former general manager of Fudian Bank during the Republic of China. It is a two-storey brick-and-wood structure, similar to Villa No. 8 Houxin Street, with a blue brick exterior and yellow wiring. However, unlike the Villa No. 8 Houxin Street, the building's graphic design is more chic and adopts a polygonal shape. The entrance of the villa is cleverly designed on the concave balcony on the second floor, which acts as a wind porch and shows the ingenuity of the functional design.

7.3 后新街 7 号别墅 2

建筑坐北朝南，米黄色抹灰的外墙上，绿色格子窗户点缀其间。二楼设有长长的阳台，为整栋建筑的点睛之笔。

7.4 后新街 8 号别墅 3

沿街而建，坐西朝东，建筑风格简约现代。整座建筑外墙采用青砖砌筑，灰色水洗石条形窗台和圆形窗套与青绿色格子窗相得益彰。土黄色的走线镶嵌在墙体与屋面接缝之处，为建筑增添了一抹生动的色彩。

7.5 盘龙路 17 号别墅 4

为民国时期原富滇银行总经理的私宅。它是一座两层的砖木结构建筑，与后新街8号别墅相似，都采用了青砖外墙和黄色走线装饰。然而，与后新街8号别墅不同的是，这座建筑的平面设计更加别致，采用了多边形的形状。别墅的入口被巧妙地设计在二层的凹阳台上，既起到了遮风挡雨的风雨廊作用，也展现了功能性设计的巧思。

8

北京路

<片区>

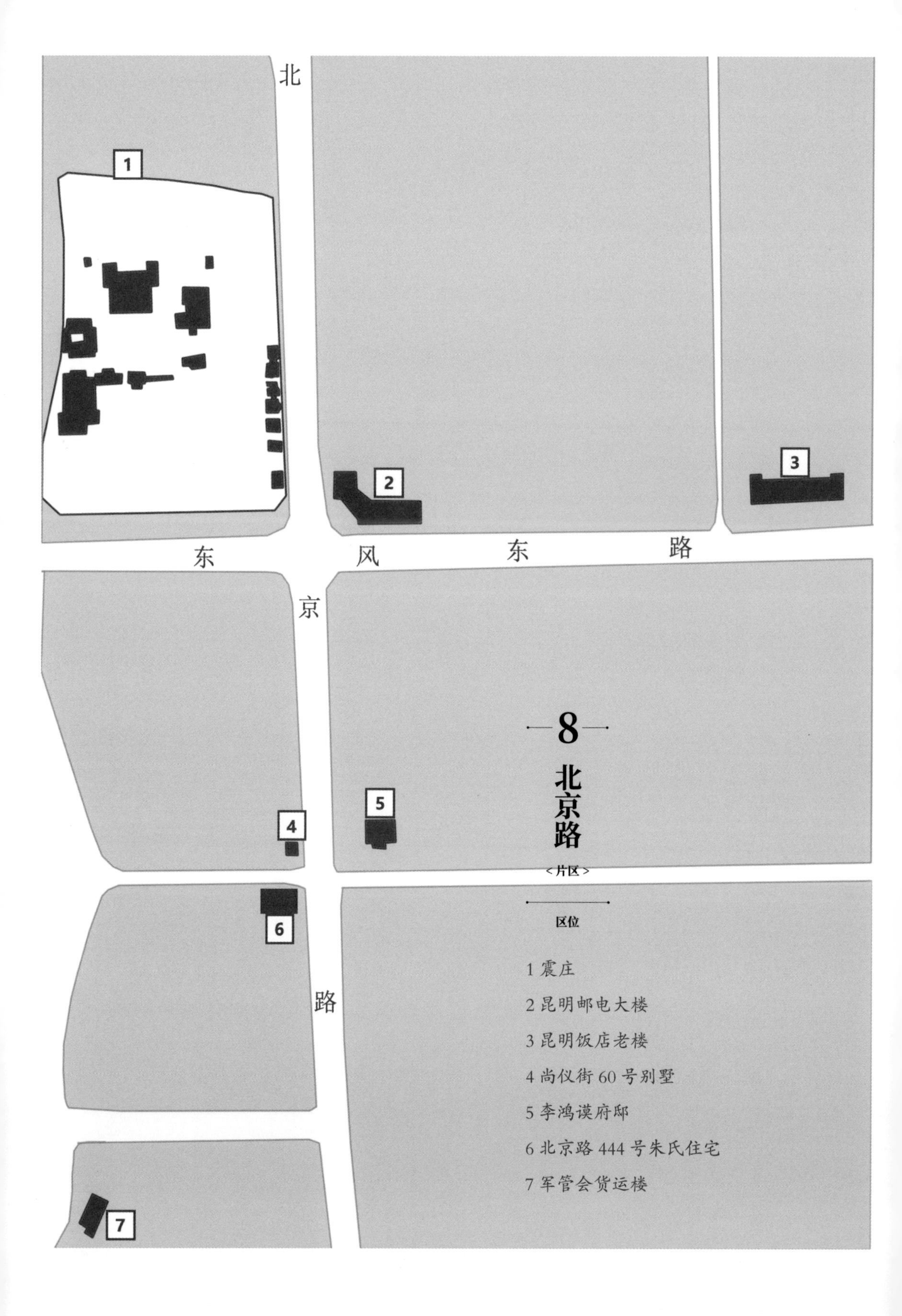
北
东
风
东
路
京
路
1
2
3
4
5
6
7
8
北京路
<片区>
区位
1 震庄
2 昆明邮电大楼
3 昆明饭店老楼
4 尚仪街60号别墅
5 李鸿谟府邸
6 北京路444号朱氏住宅
7 军管会货运楼

8 北京路片区

◈ 地名由来

北京路

北京路南起昆明火车站，北至昆明绕城高速。清代，北京路的中段由南至北，分别由塘子巷、太和巷、太和街、咸和街、福德街、穿心鼓楼组成，1937年塘子巷至穿心鼓楼一段也被称为环城东路；1950年前，北京路南段、北段多为菜地、农田和水塘。1966年贵昆铁路建成通车，铁路连通了昆明与北京，而北京路正好是贯通昆明南、北火车站的道路。故这条路被改称为北京路。

8.1 Li Hongmo's Residence (Li Residence) 5

Located in the gardens of what was then Taihe Street (Beijing Road), the Li Mansion was a two-storey, stone-built house, a blend of Chinese and French architectural styles. The house has a Chinese glazed tile roof with cantilever eaves, a French main structure and interior decoration. Surrounded by green trees and camellia flowers, the eight-sided wind gate at the corner of the building opens to Taihe Street. One of the most striking is that the seemingly uncarved, convex and uneven stones on the exterior walls are rough and natural, in sharp contrast to the clean and tall glass.

Li Hongmo (1894-1964), a native of Luliang, Yunnan Province, served as the former chief of Yunnan Police, deputy commander of Yunnan Air Defense Command, deputy commander of Kunming Security Command. Li Hongmo was known for his hospitality and hired skilled Chinese and Western chefs, making the Li residence a social venue for entertaining Yunnan military and political leaders, as well as celebrities such as Long Yun, Lu Han, and Yan Xiecheng.

8.1 李鸿谟府邸（李公馆）5

李公馆坐落在当时的太和街（北京路）的花园中，是一座石头砌就的两层洋楼，融合了中法两国的建筑风格。这座房子拥有中式的琉璃瓦和挑檐的屋顶，法式的主体结构和内部装饰。被绿树掩映、茶花簇拥下的石房子，建筑转角处的八面风大门面向太和街敞开。其中最引人注目的是，外墙上那些看似未经雕琢、凸凹不平的石头，粗犷而自然，与干净利落的高大玻璃形成鲜明的对比。

李鸿谟（1894—1964年），云南陆良人，曾任国民时期原云南省警务处处长、云南防空司令部副司令、昆明警备司令部副司令。李鸿谟以其热情好客而闻名，他同时聘请了技艺高超的中厨和西厨，使得李公馆成为宴请云南军政要员和社会名流如龙云、卢汉、严燮成等的社交场所。不仅蒋介石、宋子文等国民党政要，还有美国驻华大使约翰逊、史迪威、陈纳德等美军将领都曾在此聚会。

8.2 Zhenzhuang 1

Zhenzhuang is located at west of Beijing Road and east of Panlong River. It was built in 1936 and was the former chairman of Yunnan Prooince in the Republic of China period, Mr. Long Yun's residence.After 1949, it became the state Guesthouse of Yunnan Province to receive heads of states and important guests. In February 2023, parts of the Zhenzhuangying Hotel was opened to visitors.

From 1914 until 1919, the newly built "Qian Building" served as the German consulate, and then Longyun selected the "Qian Building" and the surrounding land, on which a private garden was built in 1936. In Zhen Zhuang are dispersely distributed Jin Building (Jin Building), embroidery Building (auditorium), dry Building, Kun Building. The names of the four buildings are abbreviated to "Jin Xiu Qian Kun", with "a beautiful future and all the best" meaning.

8.2 震庄 1

震庄（北京路514号）东接北京路，西临盘龙江，建于1936年，前身是民国时期云南省主席龙云先生公馆。1949年以后，成为云南省接待国家元首和重要宾客的国宾馆。2023年2月，震庄迎宾馆的部分区域向游人开放。

震庄占地138亩（1亩≈666.67平方米），绿化、水面达总用地的四分之三。既有亭、台、楼、榭、假山等中式私家园林建筑，又有乾楼、坤楼、喷水池和小洋楼等欧式建筑，是昆明盘龙江畔的规模最大的历史建筑群。

1914—1919年，新建的"乾楼"作为德国领事馆，其后龙云选中了"乾楼"及其周边土地，1936年在此基础上修建了私家花园。震庄之中错落分布着锦楼（瑾楼）、绣楼（礼堂）、乾楼、坤楼，这四座楼的名称首字即为"锦绣乾坤"，具有锦绣前程、事事如意的寓意。

震庄宾馆是云南省历史最长、接待规格最高的国宾馆。刘少奇、周恩来、邓小平等中央领导，英国女王伊丽莎白二世等国际政要都曾下榻在震庄宾馆的乾楼。多年来，它为云南省的礼宾接待作出了重要的贡献。

8.3 Zhu's Residence, No. 444 Beijing Road 6

Located at 444 Beijing Road, this Republic of China villa was the private residence of Zhu Ziying, the former logistics chief of the Kuomintang Security Command. It is three storeys high, with a brick structure and pitched roof, and the exterior walls are coated with pale yellow plaster and bean sand stone. The architectural design and decoration style is modern and simple, all doors and windows are made of teak material, and the floor is teak parquet. It has been repaired many times and is very well preserved.

8.3 北京路 444 号朱氏住宅 6

这座民国时期的别墅位于北京路444号，是原国民党保安司令部后勤处长朱子英的府邸。它高有三层，采用砖石结构和坡屋顶，外墙涂有浅黄的抹灰和豆沙石材料。建筑设计和装饰风格现代简洁，全部门窗都由柚木材料制成，楼地面为柚木拼花包边地板。多次进行过维修，至今保存得十分完好。

8.4 No. 60 Shangyi Street 4

No. 60 Shangyi Street is a modern-style building built during the Republic of China. The building is divided into two floors and has a masonry structure. The whole building is regular and symmetrical, and the proportion of windows and walls is balanced, highlighting the simple but beautiful design style.

8.4 尚义街 60 号别墅 4

尚义街60号别墅是建于民国年间的现代风格建筑。建筑共分为两层并采用砖石结构。整座建筑规整对称、窗墙比例均衡，彰显出朴素而不失美感的设计风格。

8.5 军管会货运楼 7

军管会货运楼建于民国时期，坐落于拓东路与盘龙江交叉口。目前，该建筑为昆明铁路局所使用。其砖石结构共三层，拥有一个地下室。在设计上，建筑的二层和三层之间装饰有凸出的横线条，窗户则采用六组成对排列的形式，整齐呈现。内部设置了走廊，对称布置了若干个办公室。

作为受现代主义风格影响的代表性建筑，军管会货运楼在昆明近现代建筑中保留得相对完好。

8.6 Kunming Post and Telecommunication Building 2

The Post and Telecommunication Building, located at the intersection of Beijing Road and East Dongfeng Road, was put into use in 1959. It is laid out along two roads at an oblique angle, presenting a long "L" shaped plane with a wide facade and shallow depth. The layout along the As the main roads in Kunming, Dongfeng Road and Beijing Road, makes the post and telecommunications building visually tall and majestic.

8.7 Kunming Hotel Old Building 3

The old building of Kunming Hotel, located at 52 East Dongfeng Road, was built in 1958. The symmetrical layout of the building, relief decorative exterior walls, window covers make it a Soviet-style architecture. As a senior hotel in Yunnan at that time, it was the only hotel with elevators and heating devices.

8.6 昆明邮电大楼 2

位于东风东路12 号的邮电大楼，于1959 年投入使用。它沿着两条道路、斜角布局，呈现出一个长条形的“L”形平面，具有宽阔的立面和较浅的进深。作为昆明市主要道路的东风路和北京路，沿街布局使得邮电大楼在视觉上显得高大雄伟。它是昆明十大公共建筑之一。

8.7 昆明饭店老楼 3

位于东风东路52 号的昆明饭店老楼建于1958 年。建筑呈对称布局，浮雕花饰的外墙、窗套使之具有苏式建筑风格。作为是当时云南高级宾馆，是唯一一家安装有电梯和采暖装置的宾馆。它是昆明十大公共建筑之一。

9 一二一大街

<片区>

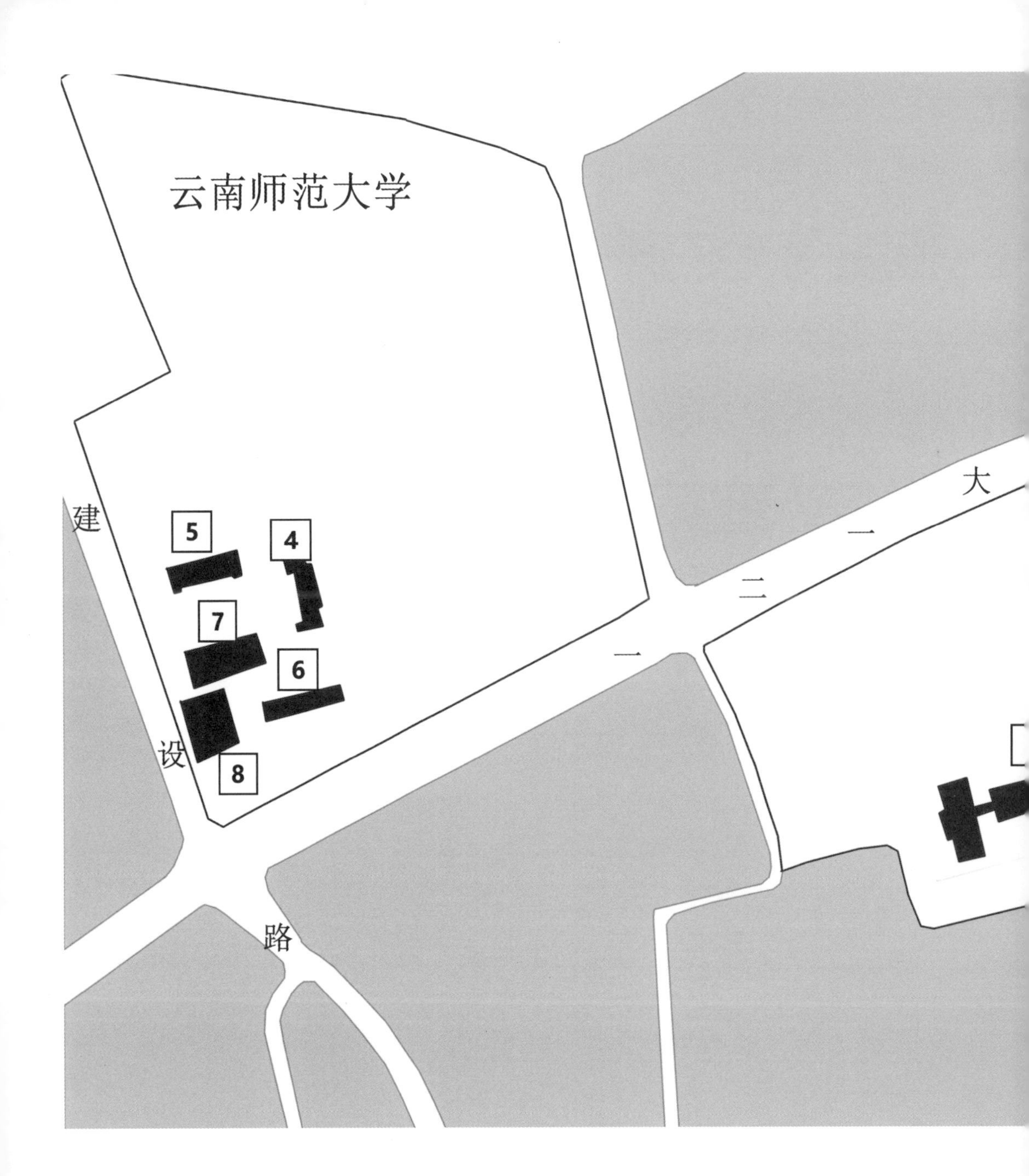
云南师范大学
建
设
路
大
一
二
一
5
4
7
6
8

1 熊庆来旧居

2 云南大学钟楼

3 云南大学生物楼、化学楼、物理楼

4 昆明师范学院物理楼

5 昆明师范学院化学楼

6 昆明师范学院第一教学楼

7 昆明师范学院第二教学楼

8 云南师范大学礼堂

9 一二一大街片区

◈ 云南大学的历史

位于昆明市五华区翠湖北路2号的云南大学原身为私立东陆大学（1923—1930年），1922年12月8日，东陆大学宣布成立。1930 年改为省立东陆大学，1932 年云南省立师范学院并入省立东陆大学（1930 —1934年）。1934年9月改称省立云南大学（1934—1938年），1937年熊庆来受聘出任云南大学校长。1938年改为“国立云南大学”（1938—1949年），1946年被《不列颠百科全书》列为中国15所在世界最具影响的大学之一。至1978年，云南大学被教育部列为全国88所重点大学之一，1997年11月正式成为国家“ 211 ” 工程首批建设的 61 所大学之一。

昆明师范学院位于五华区华山街道一二一大街，云南师范大学的前身为“国立西南联合大学（昆明师范学院）（现云南师范大学）的历史，1946年西南联大离开后，联大师范学院留昆明独立建校，定名为国立昆明师范学院”。1984年4月，经云南省人民政府批准，更名为云南师范大学。

9.1 Xiong Qinglai Old Residence 1

The former residence of Xiong Qinglai is located in a two-storey building in front of the "Examination Shed" on the east side of the Huizeyuan and the court of Yunnan University. House was built in 1937, the north to the south, is gabled roof, four wide, two deep brick and wood structure of the two-storey building. The first floor is the reception room of Mr. Xiong Qinglai, and the second floor is the bedroom or office. The main building is a two-storey building of civil structure. There is a corridor on the south side of the upper and lower floors, plain and elegant. The old house is greatly influenced by Western architecture, and has the characteristics of similar Western architecture during the Republic of China, and is well preserved. Its structure is a civil structure, the roof structure is a modern wooden frame structure, the external wall is light yellow, and the middle of the wall is decorated through the waist line.

9.1 熊庆来旧居 1

熊庆来旧居坐落于云南大学校本部会泽院和至公堂东侧，是原云南贡院考棚前的一座二层楼房。始建于1937年，坐北朝南，是一座单檐悬山顶、面阔四间、进深两间的砖木结构的两层楼房。一楼为熊庆来先生的接待室，二楼为卧室、办公室。主体建筑为土木结构二层楼房，外墙为浅黄色，墙身中部做装饰腰线贯通。建筑南面设有走廊，朴素而优雅。旧居受西洋建筑影响较大，细格窗、走廊等具有民国时期西式建筑特点，保存较好。

【 **Historical background** 】

Xiong Qinglai was born on September 11, 1893 in Xizai Village, Mile, Honghe Hani and Yi Autonomous Prefecture, Yunnan. He is a mathematician, educator, pioneer of modern mathematics, and one of the main pioneers of function theory in China. From 1921 to 1936, Xiong Qinglai successively served as a professor in Yunnan Alpha Industrial School, Yunnan Road Administration School, Nanjing Southeast University, Shaanxi Northwest University, and Tsinghua University. In 1937, Xiong Qinglai became president of Yunnan University.

1

【历史背景】

1893年9月11日，熊庆来出生于云南省红河哈尼族彝族自治州弥勒市息宰村，是中国数学家、教育家、中国现代数学先驱、中国函数论的主要开拓者之一。1921年至1936年，熊庆来先后在云南甲种工业学校、云南路政学校、南京东南大学、陕西西北大学、北平清华大学任教授。1937年，熊庆来任云南大学校长。

9.2 Biology (teaching building)-Physics (teaching building)-Chemistry (teaching building) Building 3

The Chemistry, Biology and Physics Building of Yunnan University was built in the 1950s and designed by Professor Yao Zhan of the Department of Civil Engineering of Yunnan University. The physical building is a three-storey Soviet-style building with brick-concrete structure from south to north. There are six round columns about 30 meters high on the front steps, and the column head sculpture is simple. There are double cylindrical cross-corridors, connecting with the back building, and the whole building is majestic. Biology building and Chemistry building are divided into east and west two floors, between the two floors for the physics museum. The east building sits west to east and west building sits east to west. They are three-storey brick structure Soviet-style architecture that, east and west building front steps are standing six circular columns, magnificent. The three buildings are perfect, showing the shape of Chinese character "I". The whole architechture is tall, of magnificent volume, dignified and elegant.

9.2 云南大学生物楼、物理楼、化学楼 3

云南大学化学、生物、物理楼建于20世纪50年代，由云南大学土木系姚瞻教授主持设计。物理楼坐南向北为三层砖混结构苏式建筑，正面台阶上立有高约30米的6根圆形立柱，柱头雕塑简洁。与后楼连接处，建有双圆柱式跨廊，整幢大楼雄伟庄严。生物楼和化学楼分为东西两楼，两楼之间为物理馆，中间有廊道连接，东楼坐西向东，西楼坐东向西，均为三层砖石结构苏式建筑，东、西楼正面台阶上均立有6根圆形立柱，雄伟壮观。三楼珠联璧合，呈“工”字造型，整个建筑物高大、体量宏伟，端庄典雅。教学楼竣工后一直担负着云南大学理科的教学、学习、试验等基地的任务，是培养生物教学人才的集散地。为典型的西式风格建筑物，有一定的历史研究价值和艺术价值。

9.3 Bell Tower in Yunnan University 2

After the completion of the science experimental building of Yunnan University, the auxiliary engineering water tower was built as a bell tower. Located to the west of Huize Building in the East Land Park of Yunnan University, which was built in 1956. The tower has 7 floors, 26 meters high, and the steel frame at the top of the tower is 30 meters high. It was designed by Professor Yao Zhan, Director of the Department of Civil Engineering, Yunnan University.

9.3 云南大学钟楼 2

云南大学理科实验楼建成后，兴建配套工程水塔兼作钟楼。云南大学钟楼位于云南大学东陆园会泽楼西面，始建于1956年。塔共7层，高26米，连塔顶钢架共高30米。由云南大学土木系主任姚瞻教授设计。

为方形碉楼式建筑，共7层，内有楼梯通至顶层，第7层兼作水池，把地下水通过水泵压至水池，再通过管道连到各幢建筑。云南大学内的钟楼，半个世纪以来，至今还在按北京时间准点响钟，音响千米。成为大学内的一道景观，具有一定的历史、科学、艺术价值。

9.4 Physics Building, Kunming Normal University 4

The Physics Building of Kunming Normal College was built in the early years of the founding of the People's Republic of China and began construction in 1956. The Physics Building has a history of 70 years with the development of the school. First, it has witnessed the development of Teachers College Southwest Associated University–National Kunming Teachers College–Kunming Teachers College–Yunnan Normal University, and has important historical and cultural significance. Second, it reflects the construction design style in the early days of the founding of the People's Republic and is typical. The layout of the building is symmetrical in the left and right axis, the plane is regular, the middle is high on both sides of the low, the main building is equipped with an observatory, and the back is wide and slowly extended: the secondary is a "one-section" structure—including the eaves, the wall body, and the feet of the 3 parts.

4

9.4 昆明师范学院物理楼 4

昆明师范学院物理楼，于1956年开始建设，物理楼伴随着学校的发展已有70年的历史。一是物理楼见证了西南联大师范学院、国立昆明师范学院、昆明师范学院、云南师范大学的发展历程，具有重要的纪念、教育等历史文化意义。二是反映了中华人民共和国成立初期的建筑设计风格，具有典型性。建筑布局为左右呈中轴对称，平面规整，中间高两边低，主楼设有天文台，“一段式”结构——包括檐部、墙身、勒脚3个部分。三是在建筑材料、结构、施工技术的建筑工程技术和科技水平，体现一定的科学技术价值。完整保留了1956年建设之初全套设计和施工蓝图图纸，详细记录了物理楼建造工艺、材料、建设细节等是极其珍贵的资料，完整清晰地反映了中华人民共和国成立初期的建筑风格。

9.5 Chemistry Building, Kunming Normal University 5

The construction of Chemistry Building of Kunming Normal College began in 1952, which has been accompanied by the development of the school for 70 years. The layout of the building is symmetrical, and the three-section facade design has typical characteristics of the times.

9.6 No. 1 Teaching Building of Kunming Normal University 6

The construction of the first teaching building of Kunming Normal College began in 1954 and has been accompanied by the development of the school for 70 years. The layout of the building is symmetrical in the left and right axis, the plane is regular, and the corridor is wide and slowly extended; secondly, there is a "three-stage" structure—including the eaves, the wall body, and the feet of the three parts.

9.5 昆明师范学院化学楼 5

昆明师范学院化学楼建于1952年，建筑平面布局呈中轴对称，三段式立面设计具有典型的时代特征。

9.6 昆明师范学院第一教学楼 6

昆明师范学院第一教学楼建于1954年，建筑布局为左右呈中轴对称，平面规整，中间高两边低，回廊宽缓伸展；其次是有“三段式”结构——包括檐部、墙身、勒脚3个部分。

9.7 昆明师范学院第二教学楼 7

昆明师范学院第二教学楼建于1954年，建筑平面布局呈中轴对称，三段式立面设计具有典型的时代特征；同时完整保留了施工蓝图，详细记录了建筑建造工艺、材料结构、建设细节等信息，完整清晰地反映了中华人民共和国成立初期的建筑风格。

9.8 云南师范大学礼堂 8

云南师范大学礼堂是云南师范大学正式更名以来排名前列的一座礼堂。其平面布局、外立面形式、建筑材料、结构工艺等是20世纪80年代大型公共建筑的典型体现。

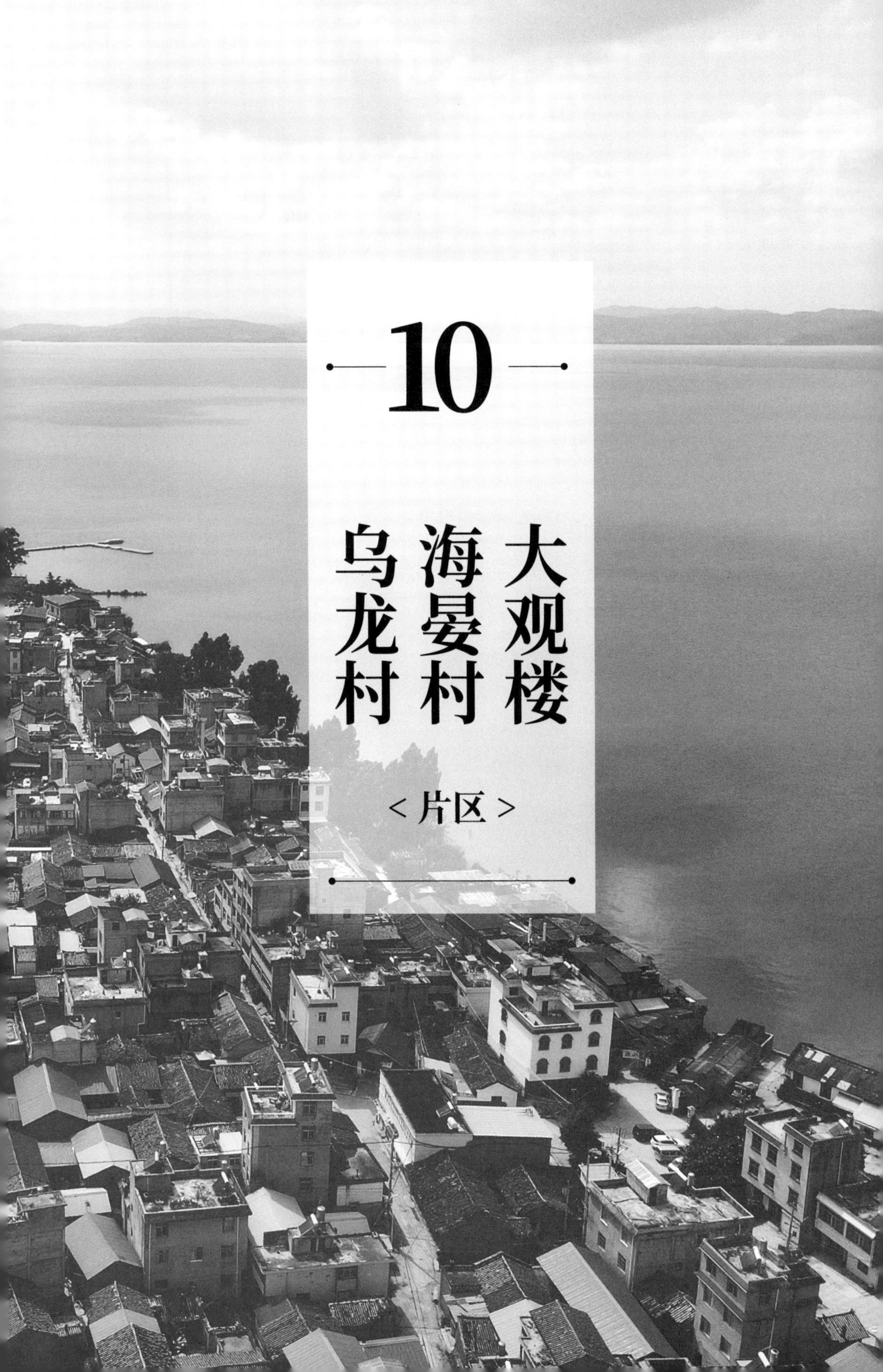

10

大观楼 海晏村 乌龙村

< 片区 >

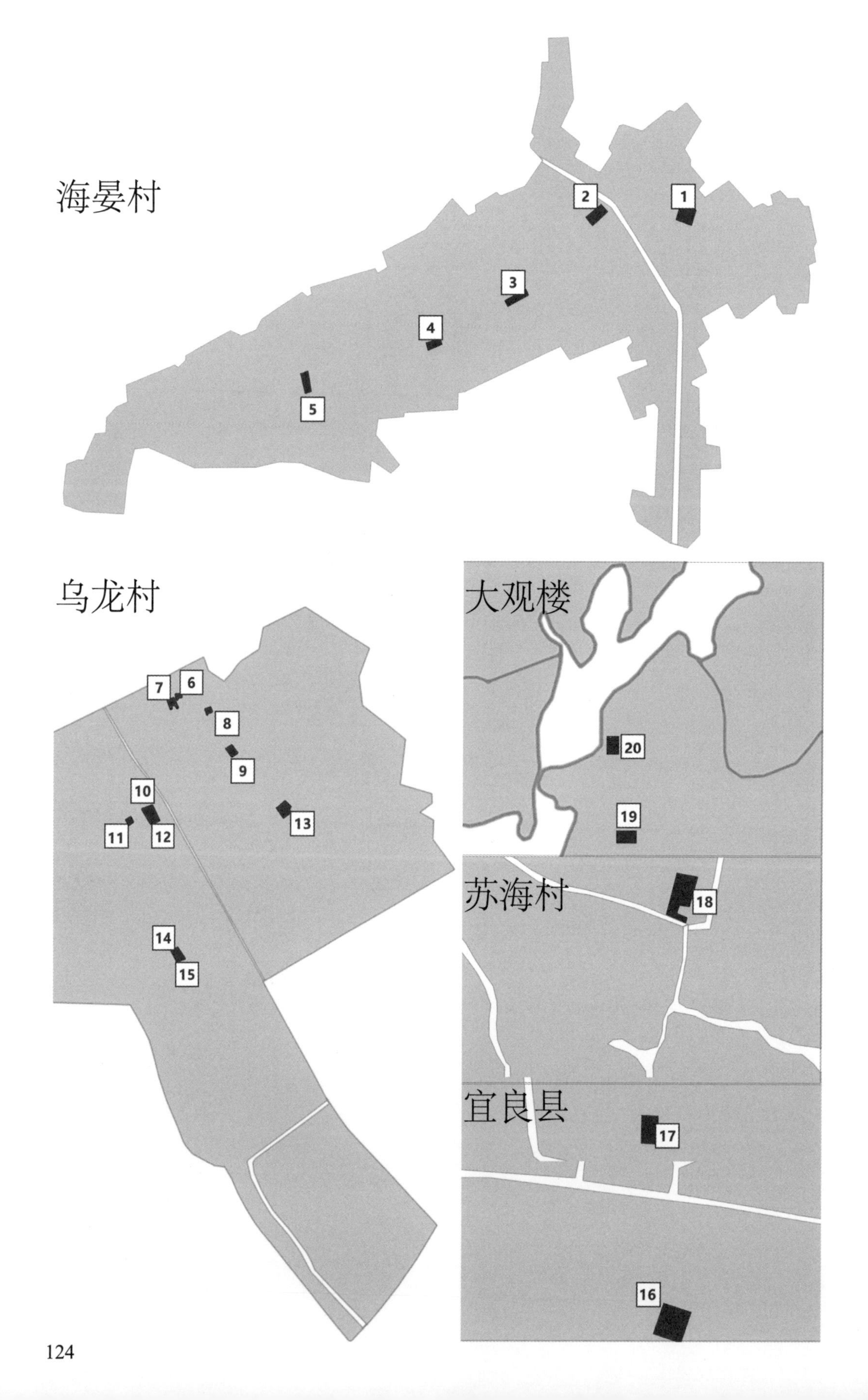
海晏村
1
2
3
4
5
乌龙村
7
6
8
9
10
11
12
13
14
15
大观楼
20
19
苏海村
18
宜良县
17
16

—10—

乌龙村 海晏村 大观楼

<片区>

区位

1 海晏村关圣宫

2 海晏村吕祖阁

3 海晏村 488 号民居

4 海晏村 615 号民居

5 海晏村西寨门、海晏村老客事房

6 乌龙村民居 203 号

7 乌龙村民居 204 号

8 乌龙村民居 221 号

9 乌龙村民居 187 号

10 乌龙村民居 80 号

11 乌龙村民居 40 号

12 乌龙村民居 81 号

13 乌龙村民居 145 号

14 乌龙村民居 63 号

15 乌龙村民居 64 号

16 宜良县何家大院

17 宜良县祝氏民居

18 苏海村苏子鹄旧居

19 鲁氏别墅

20 杨家大院

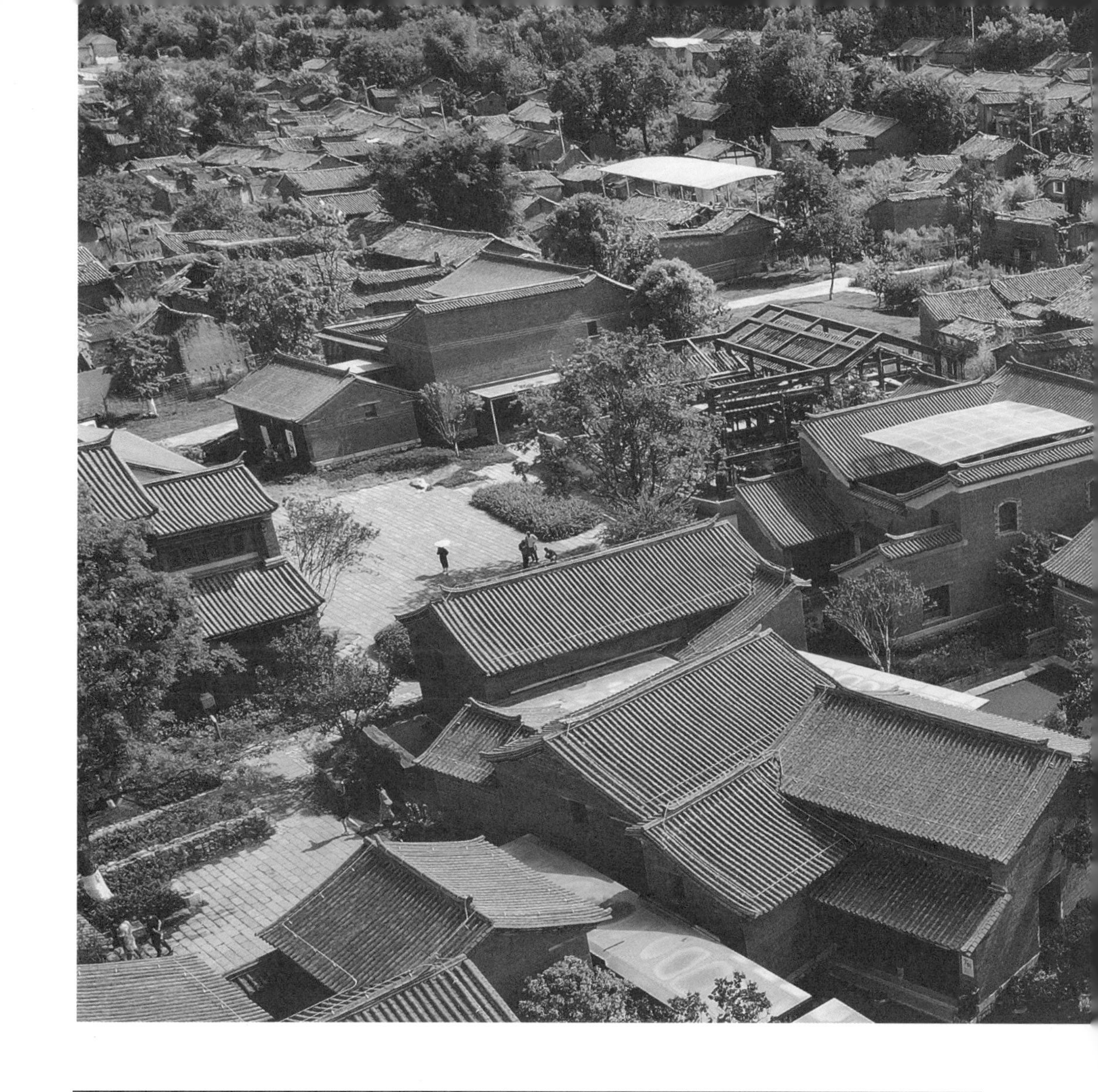

10 大观楼、海晏村、乌龙村片区

大观楼

大观楼公园，古称近华浦，在昆明城西南，明初沐英滇池草海北滨建“西园”。

清康熙二十一年（1682年），湖北僧人乾印在此建观音寺。清康熙三十五年（1696年）由巡抚王继文修建，题名“大观楼”并建了观稼堂、催耕馆、揽胜阁、涌月亭等。孙髯翁题180字长联挂于大观楼，被誉为“古今第一长联”。清代咸丰七年（1857年）楼台亭阁均遭兵燹。清代同治五年（1866年）马如龙在原址重建大观楼为三层木结构，四方形，楼顶为四角攒尖式，正中冠以高耸的宝顶，雄伟壮观。1931年，大观楼增设了假山、长堤，作为公园使用。1954年以来，扩充原址，将李园、庾园、鲁园、丁园、柏园皆纳入公园范围。

海晏村

海晏村位于滇池之滨，是为数不多的紧邻滇池且较完整地保留了云南民居特色风貌及滇池渔耕文化特点的历史村落。海晏村是滇池现存四个古渡口与古码头之一，其余三个为海口镇、官渡古镇、河泊村。早在明清时期，粮草通过水上交通，在海晏码头中转，供给内陆各个城镇，当时的海晏村已成为滇池东岸与南岸的重要贸易码头。

乌龙村

滇池东岸的乌龙村是昆明环滇池地区保护较完整的村落，至今已有600多年历史，古渔村“山居宛卧龙，海近乃曰浦”。村落处于滇池岸边的天然港湾，傍晚归港的船只众多，船上村中灯火如繁星，形成老呈贡八景之一“渔浦星灯”的盛景。

10.1 Lu's Villa 19

Lu's Villa, also known as Ziquan Pavilion, is located in the south garden of Kunming Daguan Park, built in 1927 as Lu Daoyuan's private garden villa. The environment of Lu's Villa is beautiful and the vision is wide, you can see Dianchi Lake close, you can also overlook the Western Mountains. It has a huge private garden, the layout of the whole garden is small and unique, winding paths, quiet and elegant. stone boats and bridges, lakes and pools of green trees in the garden, constitute a picture of contentment. Lu Daoyuan (1900-1985), born in Changning County, Yunnan, graduated from Yunnan Army Military School.

10.1 鲁氏别墅 19

鲁氏别墅，又称子泉别业馆，位于昆明市大观公园南园内，建于1927年，为鲁道源的私家园林别墅。鲁氏别墅的环境优美，视野开阔，可近观滇池，也可远眺西山。它拥有一个巨大的私家园林，整个园林的布局小巧别致，曲径通幽，宁静典雅。园中的亭台楼阁、石舫曲桥、湖池绿树，构成了一幅怡然自得的画面。

该别墅的平面布局呈“L”形，墙基、墙角、窗套和门套都采用石砌的方式，屋顶则是平瓦屋面。

整座建筑是单层的欧式建筑。别墅前面有一个月台，后面的石桥与圆形观景台相连接。鲁道源（1900—1985年），字子泉，云南昌宁县人，云南陆军讲武堂毕业。抗战时期，任国民党第五十八军新十一师师长、副军长，1942年任军长，曾参加长沙会战及长衡、赣东等战役。1945年抗战胜利，作为抗日名将担任南昌受降主官接受日军投降。1949年任国民党第十一兵团司令。

10.2 Yang Family Compound 20

No. 3, North Minzucun Road, is Yang Family Compound, which is composed of 4 Dali Bai-style residential buildings. The courtyard pattern is complete, the architectural workmanship is fine, and there are European architectural symbol elements.This complex was originally located in Jingding Lane of Chongren Street in the old city of Kunming, which was the residence of Dali merchant Yang's family, and also the location of Yunchangxiang Business House. In the reconstruction of the old city of Kunming, in order to protect its whole, the Yangjia compound was moved to the present position. This successful case of overall migration protection not only has historical value, but also has certain research value.

10.2 杨家大院 20

民族村北路3号为杨家大院，是有大理白族风格的4座民居建筑而构成，院落格局完整，建筑做工精细，带有欧式建筑符号元素。

这座建筑群最初位于昆明老城的崇仁街静定巷，是大理商人杨家的宅院，也是云昌祥商号的所在地。在昆明旧城改造中，为了保护其整体，将杨家大院迁移至现在的位置。这一整体迁移保护的成功案例，不仅具有历史价值，还具备一定的研究价值。

10.3 Guansheng Palace, Haiyan Village 1

This building is the place of worship for Emperor Guan in Haiyan Village, and it is an important public activity place of Haiyan Village and an important witness of the historical change of the Village. The overall spatial pattern of the building is well preserved, and the building is a civil structure, which has important historical and social value.

10.4 Lyu Zu Pavilion, Haiyan Village 2

This building is the place of worship of Lyu Dongbin in Haiyan Village, which belongs to Taoism architecture, reflecting the coexistence of multi-culture in the Village and the area around Dianchi Lake. The overall spatial pattern of the building is well preserved, and the building is a civil structure, which is an important public activity place in Haiyan Village.

10.3 海晏村关圣宫 1

该建筑为海晏村祭祀关帝之处，是海晏村重要的公共活动场所、也是海晏村历史变迁的重要见证。建筑整体空间格局保存完好，建筑为土木结构，具有重要的历史、社会价值。

10.4 海晏村吕祖阁 2

该建筑为海晏村祭祀吕洞宾之处，属于道教建筑，体现了海晏村及环滇池地区多元文化的共融共生。建筑整体空间格局保存完好，建筑为土木结构，是海晏村重要的公共活动场所。

10.5 The Weet Gate of Haiyan Village 5

The West Gate of Haiyan Village is an important defining factor for dividing the inner and outer space of Haiyan Village, which plays an important role in researching the overall spatial form and history of the Village. The whole village gate is complete and is a civil structure.

10.6 Haiyan Village Old Guest Room 5

The building is connected to the village's West Gate and adjacent to Dianchi Lake, with a complete overall shape. It is an important public activity place of Haiyan Village and an important witness of the urban changes of Haiyan Village and the region around Dianchi Lake. The structure of the building is the external adobe wall of the lifting beam wooden frame, and the overall wooden frame is preserved intact with outstanding features.

10.5 海晏村西寨门 5

海晏村西寨门是划分海晏村内外空间的重要限定要素，对考证海晏村整体空间形态与历史具有重要的作用。寨门整体完整，为土木结构。

10.6 海晏村老客事房 5

该建筑与村庄西寨门相连，且紧临滇池，整体型制完整，是海晏村重要的公共活动场所，也是海晏村及环滇池区域城市变迁的重要见证。建筑结构为抬梁式木构架外包土坯墙，整体木构架保存完整，特色突出。

10.7 No. 488, Haiyan Village 3

The building evolved from "one-seal", consisting of two basically the same size "three-room, four-wing, inverted eight-foot" courtyard houses. Wood work, brick work, stone work with good craftsmanship, carving exquisite, partial painting with color, local characteristics prominent. It has important material value for studying the history of Chenggong of Ming and Qing Dynasties and enriching the history and culture of Chenggong.

10.8 No. 615, Haiyan Village 4

The building is a typical one-seal residential house, which better reflects the courtyard space of "three-room, four-wing, inverted eight-foot". The wood is beautifully carved and the local characteristics are prominent. It has important material value to enrich Chenggong's history and culture.

10.7 海晏村 488 号民居 3

该建筑由"一颗印"演变而来，由两个大小基本一致的"三间四耳倒八尺"四合院民居构成。木作、砖作、石作精工细作，雕刻精美，局部绘有彩画，地方特色突出，对研究明清呈贡历史具有重要价值。

10.8 海晏村 615 号民居 4

该建筑为典型的"一颗印"民居，较好地体现"三间四耳倒八尺"的四合院院落空间。其木作雕刻精美，地方特色突出。

10.12 No. 81.and No. 80.Wulong Village 12

The two buildings are traditional "one seal" ones, the quadrangle courtyard intact There is a pomegranate tree planted in the courtyard, which forms a good landscape. The buildings have a wooden frame, with the external walls being traditional adobe ones, and the traditional "one seal" building's green tile roof.

10.9 乌龙村民居 40 号 11

该建筑为传统“一颗印”建筑，三合院。建筑整体格局、院落型制较为清晰，与环境有机融合，较好地体现了山地建筑特色。

10.10 乌龙村民居 63 号、64 号 14 15

两栋建筑分别由两栋三合院组成，两栋建筑紧紧相连，形成独特的风貌。

10.11 乌龙村民居 80 号 10

10.12 乌龙村民居 81 号 12

两栋建筑为传统的“一颗印”建筑，四合院院落保存完整，院中栽有一棵石榴树，枝叶繁茂，形成良好的院内景观。建筑为穿斗木构架建筑，外墙为传统土坯墙，青瓦屋面的传统“一颗印”建筑。

10.18 Old residence of Su Zihu, Suhai Village 18

The building is located in Suhai Village, Dianyuan Street, Panlong District, which was established in the Republic of China. It has a good architectural style and is a valuable traditional residence in Yiliang County.

10.13 乌龙村民居 187 号 9　10.14 乌龙村民居 203 号 6

10.15 乌龙村民居 204 号 7　10.16 乌龙村民居 221 号 8

10.17 乌龙村民居 145 号 13

乌龙村的这几栋历史建筑为昆明传统“一颗印”四合院民居，建筑均保存完好，内部石作、木作精美，建筑与环境相协调。

10.18 苏海村苏子鹄旧居 18

该建筑位于盘龙区滇源街道苏海村，最早建立于民国，建筑风貌良好，是较为有价值的传统民居。

10.19 He family Compound in Yiliang County 16

No. 23, Jinguang Street, Yiliang County. The building was built by He Shouxin of Yiliang County during the Republic of China period, sitting south to north, lifting beam frame, hard peak. The whole building is composed of foyer, panel wall, main hall, side room, main hall ear room and front hall wing room, one main patio and four auxiliary patios. It has relatively complete scale, fine small wood, door frame, lintels, columns and eaves, etc., with exquisite and detailed carving, the gold column foundation in the main hall Ming room is beautifully carved, and the second room and other eave column foundations are of square-vase shape, all of which maintain the original appearance. The large scale, complete building regulations, well preserved and comprehensive residential buildings of He Family Compound have high historical architectural artistic value.

10.19 宜良县何家大院 16

何家大院位于宜良县覲光街23号。该建筑为民国时期宜良县团总何守信所建，坐南朝北，四合五天井式布局，抬梁式构架，硬山顶。整栋建筑由门厅、板壁、正堂、厢房、正堂耳房和前厅耳房，一个主天井和四个副天井构成，规模较齐备，小木作精良，门框，门楣、柱枋、檐板等具有精巧细致的雕刻，正厅明间金柱柱础雕刻精美大气，次间及其他檐柱柱础为四方宝瓶形，均保持原貌。何家大院建筑规模较大，建筑规制较全，是保存状况较好、较全面的民居建筑，有一定的历史价值和艺术价值。

10.20 Zhu House in Yiliang County 17

No. 47 Xuepo, Kuangshan Street, Yiliang. The building was constructed by Fan Shisheng, a general of the Dian Army during the Republic of China. The building is located in the south facing the north, backed by the Jacana Hill, quadrangle-style layout, lift-beam frame, hard hilltop, blue and gray bois, mainly composed of the screen wall, foyer, atrium, left and right wing rooms and patio, except for the vestibule, are all two-storeyed. The atrium has a stone base, on both sides of the foot and the side room corridor connection, part of the wood carving. In addition to the damaged lighting walls, the rest of the existing buildings have maintained the original layout and architectural style, which has a certain historical and artistic value.

10.20 宜良县祝氏民居 17

祝氏民居位于宜良匡山街学坡47号。该建筑为民国年间滇军将领范石生所建。建筑坐南朝北，背靠雉山，四合院式布局，抬梁式构架，硬山顶，青灰布瓦，主要由照壁、门厅、正厅、左右厢房和天井构成，除前厅外，均为二层，“一颗印”式格局，正厅有条石基座，两侧有踏跺和厢房走廊连接，部分木作有雕刻，现存建筑除照壁已损毁外，其余均保持原布局和建筑风貌，具有一定历史、艺术价值。

后 记

——时光为眸 建筑之美

《昆明历史建筑》采用中英双语、地图导览、图文并茂的形式，通过 178 张建筑照片、22 张建筑测绘图纸，以“文明街”“南强街”“翠湖”等 10 个片区为引领，带领读者跨越时空的界线，纵览春城各处历史建筑的价值特点、保护状况，回溯历史。

全书所选建筑时跨清末至上世纪 50 年代，地域从城市到乡村，类型涵盖传统民居、别墅、园林和观览建筑等 12 种类型，对昆明历史建筑进行了系统性的分析研究，为历史建筑要素的价值评价和要素构成提供了崭新的视角。

该书由昆明理工大学原副校长，博士生导师周峰越教授带领团队，基于对昆明历史文化名城 20 余年研究，在全过程参与了昆明历史建筑的调查评估、挂牌保护、规划管理、维修实践的基础上，全面收集、整理并研究了自昆明建城一千多年保存至今的 112 处历史建筑。为了更加直观地让读者了解建筑历史、建筑区位，通过多次到实地踏勘、拍照、走访，对 112 处建筑划定分布位置、总结特点、整理不同时期的照片、建筑测绘图件，对建筑的特点全面梳理。

2010年，本书主编周峰越与瑞士历史建筑专家温纳·施图兹博士、住建部督察员丁秀云在昆明呈贡文庙指导修复工作（左一温纳·施图兹、中间周峰越、右一丁秀云）

2005年，本书主编周峰越向第二批历史建筑（即保护建筑）光华街63-77号产权人授牌

昆明历史建筑保护简况

1982年，昆明市与瑞士苏黎世市结为友好城市。从 1996 年开始，老城保护项目成为两市城市规划技术合作的内容。1996 至 2018 年的 22 年间，昆明与苏黎世开展的老城保护项目，共同调查了市域传统村镇、文明街片区、东寺街片区、祥云片区、近现代建筑等。在项目成果的基础上，昆明市政府开展一系列遗产保护工作：一是，保留了文明街片区，整合了东寺街与书林街片区功能，恢复金马碧鸡坊，打通东、西寺塔连廊等；二是，挂牌保护了代表昆明不同时期特色的第一、二批历史建筑；三是，促成 2000 年昆明市规划局成立昆明历史遗产保护机构——昆明历史街区与建筑保护办公室（即“保护办”）。该团队的 3 名成员是保护办的主要工作人员，周峰越在昆明工作期间，作为保护办主任，带领高雪梅、金浩萍积极推动并开展了历史建筑挂牌保护、规划管理工作。

保护办成立后首先开展的工作就是组织开展对昆明文明街历史街区的 178处历史建筑普查，首次初步建立了昆明历史建筑基本情况档案，提出了“第一批登录历史建筑侯选名单”。昆明市政府批准的第一批历史建筑于 2002 年 9 月正式挂牌保护。至 2022 年的 20 年间，昆明市政府已批准了 6 批历史建筑 112 处，其中的 46 处建筑因保护状况较好已升级为文物保护单位。

编撰出版本书的目的、意义

编撰出版《昆明历史建筑》可以更加全面地展示昆明历史建筑的保护与管理，在展现昆明历史建筑特色、特点的同时，有效地提高文化遗产传承水平，让更多读者、游客、市民参与到历史文化的传承中来，清晰地了解历史建筑的位置及背后的故事，让文化遗产找得到、看得见、感受得到。

此书具体呈现了昆明历史建筑的特点：一是建筑分布区域广，主要布在滇池北岸、东岸地区，以历史文化街区和历史地段 42 处（占比 37.50%）、道路沿线27 处（占比 24.11%）居多。二是建筑始建时间跨度大，以近代、现代建筑为主，建成 75 年以上的建筑有 88 处，占比 78.57%。三是建筑类型差异大，占比前三的是，传统民居 59 处（52.68%）、近代别墅 14 处（12.50%）、商业建筑 11 处（9.82%）。

通过此书，让大家重新认知了历史建筑保护体系价值，对历史建筑现状情况的进行了系统性、网络性分析研究，对历史建筑要素的价值评价和要素构成提供了新的角度。

昆明历史建筑保护简况

一是建立了昆明历史建筑查询检索体系。通过对历史建筑的准确定位、逐一标识、索引编号，完整地建立了昆明历史建筑查询检索体系，为进一步研究建筑历史、城市变迁、历史建筑的选取和判定提供了技术支撑。

二是昆明历史建筑保护价值的完整提炼。历史建筑是在不同时间节点、不同地点、不同参与人员，多批次进行调查评估、批准公布，对其建筑保护价值的提炼存在标准不一致、内容不全面等问题。通过查阅大量史料、文件，对照国家相关技术规范和标准，对昆明已公布的所有历史建筑的保护价值，进行再次评估和完整提炼，以期对今后历史建筑的调查研究、价值评估提供学术参考。

三是为今后扩展性研究提供了基础资料。完整全面整理了已公布昆明历史建筑的位置、历史背景、价值特色、维修成效等，相关研究学者可通过此成果，开展历史建筑保护、管理的扩展性研究。

2005年，昆明市历史建筑保护办公室到文明街指导修复工作（左一本书副主编高雪梅、右一本书主编周峰越）

2008年，昆明市历史建筑保护办公室到文明街指导修复工作（左一本书主编周峰越、右一本书副主编金浩萍）

2022年，本书摄影段文调研云南历史建筑（左一段文）

2023年，本书摄影刘伶俐调研文明街（中间刘伶俐）

昆明历史建筑一览表

1	马家大院（小银柜巷 7 号）	39	原云南省博物馆（五一路 118 号）	77	陈氏民居（晋宁打黑村）
2	文明街小银柜巷 8 号宅院	40	邮电大楼（东风路东路 52 号）	78	祝氏民居（宜良匡山街学坡 47 号）
3	懋庐（景星街吉祥巷 18、19 号）	41	昆明饭店老楼（东风路东路 52 号）	79	何家大院（宜良觐光街 23 号）
4	王炽故居（文庙直街 103 号）	42	南屏电影院（晓东街 2 号）	80	景星街 92–94 号宅院
5	藜光庐（文庙东巷 5 号）	43	飞虎队招待所（昆瑞路 38 号）	81	景星街 30 号宅院
6	文明街幸福巷 1 号宅院	44	核工业商务楼（东风西路 176–182 号）	82	光华街 46–50 号宅院
7	文明街幸福巷 4 号宅院	45	宝善街 171–173 号宅院	83	文庙直街 78 号宅院
8	文明街幸福巷 5 号宅院	46	宝善街 175–177 号宅院	84	宝善街 193 号宅院
9	文明街幸福巷 6 号宅院	47	飞虎队俱乐部（宝善街 179 号）	85	北后街 27–30 号宅院
10	欧式宅院（文明街 11 号）	48	祥云街 43–44 号宅院	86	北后街 31–33 号宅院
11	文明街 16 号宅院	49	祥云街 45–48 号宅院	87	北后街 34 号宅院
12	文明街 22 号宅院	50	李鸿谟府邸（北京路茶花公园内）	88	端仁巷 17 号宅院
13	文明街 28 号宅院	51	震庄迎宾馆别墅	89	端仁巷 18 号宅院
14	甬道街西卷洞巷 1 号宅院	52	北京路 444 号住宅	90	中国科学院(昆明分院)办公楼(护国路)
15	正义路四通巷 2 号宅院	53	尚义街 60 号住宅	91	昆明师范学院物理楼（一二一大街）
16	正义路四通巷 3 号宅院	54	后新街 7 号住宅	92	昆明师范学院化学楼（一二一大街）
17	正义路居仁巷 8 号宅院	55	后新街 8 号住宅	93	昆明师范学院第一教学楼(一二一大街)
18	傅式宅院（居仁巷 10 号）	56	盘龙路 17 号住宅	94	昆明师范学院第二教学楼(一二一大街)
19	正义路居仁巷 11 号宅院	57	甘美医院（青年路 504 号）	95	云南师范大学礼堂
20	马氏宅院（花椒巷 5 号）	58	熊庆来故居（云南大学）	96	缅甸战役中国阵亡将士纪念碑
21	崇仁街 3 号中宅院	59	生物楼（云南大学教学楼）	97	唐家花园遗址（昆明动物园内）
22	五华区崇仁街 7 号宅院	60	物理楼（云南大学教学楼）	98	昆明动物园大门
23	传统民居（省第一人民医院内）	61	钟楼（云南大学）	99	昆明动物园长臂猿馆
24	得意居（金马碧鸡商城内）	62	鲁道源私宅（鲁家花园内）	100	昆明动物园热带猴馆
25	东苑别墅（华山东路 7、8、9 号）	63	光华街 33 号宅院	101	昆明动物园金丝猴馆
26	黄河巷 37 号宅院	64	光华街 38–44 号宅院	102	昆明动物园亚洲象馆
27	石屏会馆（中和巷 24 号）	65	光华街 45—51 号	103	乌龙村民居 40 号
28	何氏宅院（富春街 83 号）	66	南屏街 68–75 号	104	乌龙村民居 63 号
29	富春街 3 号宅院	67	国防剧院（五一路 68 号）	105	乌龙村民居 64 号
30	近代别墅（原昆明市委 3 号楼）	68	杨家大院（民族村北路 3 号）	106	乌龙村民居 80 号
31	袁嘉谷故居	69	原军管会货运楼(拓东路与盘龙江交叉口)	107	乌龙村民居 81 号
32	翠湖北路 3 号宅院	70	苏子鹄旧居（苏海村）	108	乌龙村民居 187 号
33	卢汉公馆（翠湖南路 4 号）	71	海晏村关圣宫	109	乌龙村民居 203 号
34	光华街 63–77 号宅院	72	海晏村吕祖阁	110	乌龙村民居 204 号
35	酒杯楼东楼（光华街、云瑞东路）	73	海晏村西寨门	111	乌龙村民居 221 号
36	酒杯楼西楼（光华街、云瑞西路）	74	海晏村老客事房	112	乌龙村民居 145 号
37	原云南省科技馆（翠湖西路 1 号）	75	海晏村 488 号民居		
38	云南省艺术剧院（东风路西路 138 号）	76	海晏村 615 号民居		

注明：表中如棕色字体的马家大院等 46 处历史建筑，现已升级列为文物保护单位。

勘校说明：在出版之时，将书中所有彩色照片统一调整为黑白色，导致如北后街端仁巷 18 号宅院等建筑照片的图面效果不尽人意，望读者予以谅解。居仁巷 8 号至 11 号这组民居建筑存在建筑名称与测绘图纸不对应之处，请古建筑研究者以规划主管部门存档备案的建筑测绘图纸为准。